MECHANICAL DRAWING

TECHNIQUE AND WORKING METHODS

FOR TECHNICAL STUDENTS

BY

CHARLES L. ADAMS

ASSOCIATE PROFESSOR OF DRAWING AND DESCRIPTIVE GEOMETRY
IN THE MASSACHUSETTS INSTITUTE OF TECHNOLOGY

FIRST EDITION

BOSTON
GEO. H. ELLIS CO., PRINTERS, 272 CONGRESS STREET
1905

COPYRIGHT, 1905, BY CHARLES L. ADAMS

PREFACE.

A thorough preparation in drawing for a course in engineering or in architecture should include the study of descriptive geometry or the principles of representation; the training of the sense powers to give precision and facility in the technique of drawing; and instruction in practical freehand drawing. This book, treating of the second of the requirements named and of technical methods in execution, has been prepared for use in the first-year courses in drawing and descriptive geometry at the Massachusetts Institute of Technology. It is not intended as a prescribed set of exercises to be taken in the same form by all students, but rather as a collection of material sufficient to enable the teacher, by judicious selection, to lay out the work of a course, whether designed solely for educational training, or as an introduction to a particular course in engineering or architecture, or for a further specialization to meet the need of individual students.

To make up as far as possible for the inevitable loss, in large classes, of individual instruction in details — so necessary for the best results in technique — a large number of explanatory cuts has been introduced and minor processes have been fully explained. This detailed presentation, together with the subjects considered briefly in the last chapter, should, it is believed, make the book useful for reference in the student's future professional work.

The subject of projection, usually presented in text-books on mechanical drawing has been here omitted in the belief that, when a course includes descriptive geometry, it is unnecessary to give a portion of this subject under a different name. Furthermore, drawing from actual objects appears to be the best educational and practical introduction to descriptive geometry.

Special care has been given to the originals for the plates, which were drawn strictly in accordance with the directions in the practice exercises. The author desires to express his thanks to Mr. H. C. Bradley and Mr. A. T. Robinson for much valuable help with the text, and to other members of the Institute instructing staff for suggestions and criticism.

CONTENTS.

MECHANICAL DRAWING

CHAPTER I.

DRAWING INSTRUMENTS AND MATERIALS — THEIR SELECTION.

1. Mechanical Drawing — the language of engineering, architecture, and the mechanic arts — is representation based on descriptive geometry and expressed by means of drawing instruments. In the representation of objects, mechanical drawing deals primarily with actual figure and measurement, and is not in general concerned with the appearance of things.

2. The Instruments usually required for outline drawing are illustrated in Plates 1 and 2. Several special instruments and materials for brush work are shown in Plate 3. The following list represents the complete equipment for the drawing exercises here given. In procuring the instruments, it is advisable for the beginner to intrust their selection to an experienced draftsman; but, if this is impracticable, he should read Articles 3–13 before purchasing.

1 Set of Instruments.

2 Drawing Boards, $11\frac{1}{4}$ x $15\frac{3}{4}$ in. and 17 x $22\frac{1}{2}$ in.

2 T-squares, 15-inch and 21-inch, fixed head.

Rubber or Amber Triangles as follows : —

    1 45°–45°, 4-inch.

    1 45°–45°, 8 "

    1 30°–60°, 5 "

    1 30°–60°, 10 "

2 Irregular Curves (of the shapes shown in Plate 2).

1 12-inch Architect's Triangular Scale (divided into sixteenths of an inch and scales of $\frac{3}{32}$, $\frac{3}{16}$, $\frac{1}{8}$, $\frac{1}{4}$, $\frac{3}{8}$, $\frac{1}{2}$, $\frac{3}{4}$, 1, $1\frac{1}{2}$, and 3 in. to a foot).

1 Pricker.

Drawing Papers (in a strong envelope, 17 x 22½ in.), as follows : —
   3 Sheets Whatman's Half Imperial, cold pressed.
   6   "   "   "   "   hot   "
   6   "   Duplex Detail Paper (cut half imperial size).
   6   "   Tracing Cloth (cut half imperial size).
   2   "   Rowney's Roll Tracing Paper (cut half imperial size).
1 Block of White Practice Drawing Paper, 11 x 15 in., 24 sheets.
1 Dozen Thumb Tacks.
3 "Koh-i-noor" Pencils, H, 4H, and 6H.
1 6H "Koh-i-noor" Lead (for the compasses).
1 Fine File, or Sandpaper Pad (for sharpening pencils).
1 Stick India Ink, super-super, half size.
1 Slate Ink Slab, with cover.
1 Emerald Rubber.
1 Sand Rubber, $\frac{8}{16}$ x 1 x 1½ in.
1 Steel Eraser.
1 Agate Burnisher (of the form shown in Plate 2).
1 Piece Chamois Skin, size about 10 x 12 in.
1 Fine Oil Stone, 3-inch (for sharpening ruling pens).
1 Penholder and Pens, — ½ doz. each of Gillott's 303, and D. Leonard & Co.'s
   Ball Point, 521 F.
2 Camel-hair Brushes (of the sizes shown in Plate 3).
1 Water Glass and Tumbler (of the size given in Plate 3).
1 Medium Sized Sponge.
1 3-oz. Jar Drawing Board Paste.
1 Sketch Book, 7 x 8½ in., ledger paper, 100 pages.
1 6-inch Calipers.
1 2-foot Folding Rule.

3. **The Set of Instruments** (Fig. 1). The essential characteristic of a good instrument is that, when put in satisfactory working order, it remains so for a reasonable length of time,— a condition dependent upon excellence of material and

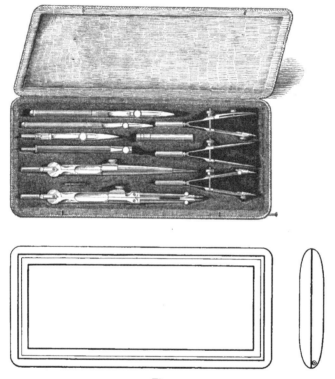

Fig. 1.

of workmanship in making the joints and tempering the points of the instrument. The difference between a good and an inferior instrument may not be easily recognized through inspection alone, as the *appearance* of the better grades is extensively imitated. To the beginner a set at three dollars may appear quite as good as one at twenty dollars. Hence his guide must be the selling price by a reputable dealer, which is from eight to twenty-eight dollars a set according to quality. For the best results in drawing, and for wear, the best instruments are necessary. A very good set, but not the best, includes: the best imitation of the Alteneder pivot-joint* compass; Alteneder-style hair-spring dividers; genuine Alteneder bow instruments and ruling pens. The price for this set (1905) should be about fifteen dollars. If the student cares to pay the difference in price, it is advisable to substitute a compass with hair-spring adjustment for that which comes with the above set.

A substitute for a case, Fig. 2, and better adapted for carrying instruments in the pocket, may be made by the student. It is of chamois leather — cloth-bound on the edges, if desired — and is fitted with pockets of the same material to hold the instruments.

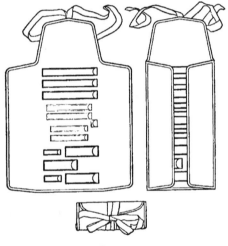

Fig. 2.

4. **The Drawing Board** (Plate 2) should be of well-seasoned, straight-grained white pine, free from sap and knots and neither shellacked nor varnished. The cleats should be of the same wood, tongued and grooved, and screwed to the board — never glued. To provide for the contraction and expansion of the board, due to atmospheric changes, the screws should pass through slots in the cleat, having a width equal to the diameter of the screw (see Plate 2). With this arrangement the board is less likely to warp or split, since, while the heads of the screws have sufficient bearing to hold the cleats in place, the slots permit the screws to move back and forth in the

* See section of the head of the Alteneder-style compass, Plate 1. Not all compasses having the handle attached to the head, as shown in the cut, are necessarily *pivot-joint* instruments.

Plate I

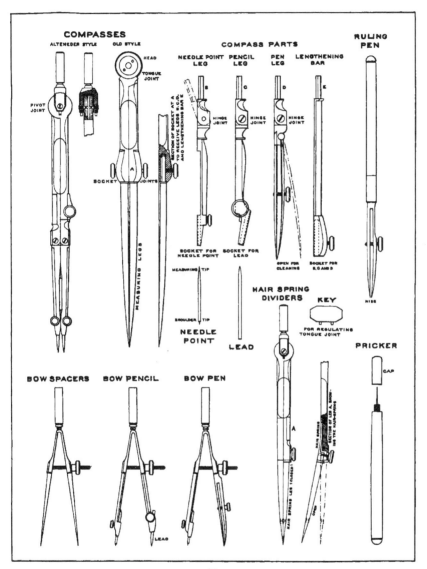

Plate 2

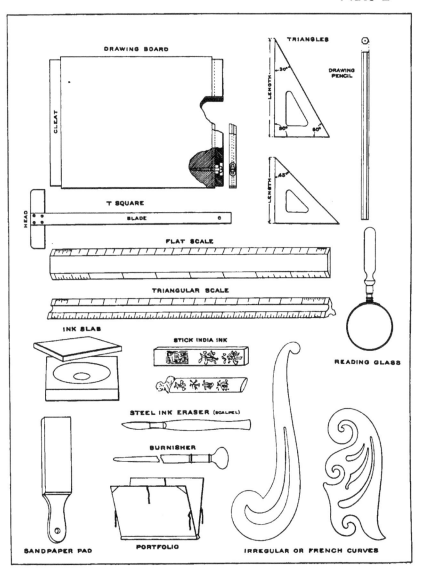

cleats with the expansion and contraction of the board. The outer edge of each cleat should be perfectly straight, and the grain of the wood parallel to the edge of the cleat. When purchasing, ask the dealer for a steel straight-edge, and with it test the straightness of the working edge of the board. Another method is as follows : —

(*a*) *To test the straightness of the working edge of a drawing board:* Place the board on a sheet of paper laid flat, and upon it, using either cleat as a straight edge, rule a very narrow line *the full length of the cleat.* Now swing the board around to reverse the ends of the same cleat with respect to the ruled line, and to bring the *same edge* of the cleat on to the ruled line. Rule again, and, if the two ruled lines coincide *throughout*, the edge of the cleat may be regarded as practically straight. The two edges of each of the two cleats should be tested.

The metal edge sometimes attached to small drawing boards is altogether unsatisfactory.

5. The T-square (Plate 2). The usual T-square is made of pear wood. A better one has the so-called ebony-lined blade. For greatest accuracy, a steel T-square and a special steel edge for the drawing board are necessary, although this T-square tends to soil the paper and to smear dry ink lines. In choosing a T-square, see that the blade is wholly free from nicks and that the grain is straight and parallel to the edges of the blade. The straightness of the blade and head should be tested according to *a*, Art. 4, or by placing each against a steel straight-edge held between the eye and the light; if the contact is perfect throughout, no light will be seen between the edges.

6. The Triangles (Plate 2) are made of wood, hard rubber, or amber. The rubber and the amber triangles are more accurate than the wooden ones. The amber is less likely to soil the drawing than the hard rubber, and also permits the lines of the drawing to be seen through the triangle — often a decided convenience. In selecting a triangle, sight across its surface to see that it is not warped. To test the straightness of the edges, proceed according to *a*, Art. 4, or place each edge against a steel straight-edge held between the eye and the light, as described in Art. 5.

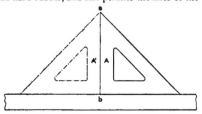

Fig. 3.

(*a*) *To test the 90° angle:* Place the triangle *A*, Fig. 3, against a straight-edge, and with a very sharp pencil draw an accurate line along the edge *ab*. If, when the triangle is turned over, as at *A'*, edge *ab* does not coincide with the ruled line, the triangle is not "square."

(*b*)   *To test the 45° angles:* After testing the T-square blade, the working

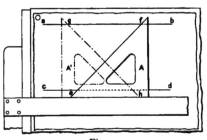

edge of the drawing board, and the right angle of the triangle, draw with the T-square two parallel lines, *ab* and *cd* (Fig. 4). With the T-square as shown and the triangle in position *A*, draw line *ef*. Bring the triangle to position *A'*, *without turning it over*, and draw *gh*. With the dividers compare the length of lines *ef* and *gh*; if they are equal, the 45° angles are correct.

Fig. 4.

(*c*)   *To test the 30° and 60° angles:* After testing the T-square blade, the working edge of the drawing board, and the right angle of the triangle, draw

with the T-square line *ab* (Fig. 5). Move the T-square down about ⅛ inch to the position shown; place the triangle in position *A*, and draw line *cd*. Turn the triangle over to position *A'*, and draw *de*. With the dividers, compare the lengths of lines *ce, cd*, and *de ;* if they are all equal, the 30° and 60° angles are correct.

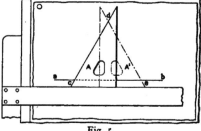

Fig. 5.

7.   The Scale (Plate 2) should be perfectly straight and free from nicks, all edges thin and sharp, and the graduations very narrow, clear-cut lines. Blunt edges and blurred graduations seriously interfere with accuracy of measurement.

8.   The Pricker (Plate 1). A good substitute can easily be made thus: Whittle out of soft, straight-grained wood a handle about 3½ inches long and tapering from ⅛ inch to $\frac{3}{16}$ inch in diameter. Break off the eye of a No. 9 sewing needle, and with a pair of pincers push the point of the needle into the smaller end of the handle, taking care to keep the needle accurately in line with the axis of the

Plate 3

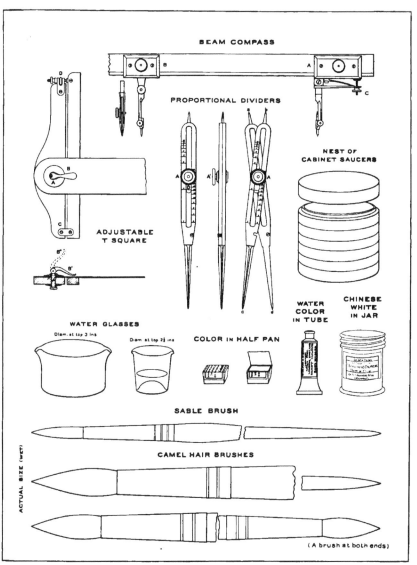

BEAM COMPASS

PROPORTIONAL DIVIDERS

ADJUSTABLE T SQUARE

NEST OF CABINET SAUCERS

WATER COLOR IN TUBE

CHINESE WHITE IN JAR

WATER GLASSES

Diam. at top 3 ins

Diam at top 2½ ins

COLOR IN HALF PAN

SABLE BRUSH

CAMEL HAIR BRUSHES

ACTUAL SIZE (WET)

(A brush at both ends)

handle. When the needle has been pushed in about three-fourths of its length, draw it out, reverse it, and force it broken end first into the hole. The needle should project about $\frac{3}{8}$ of an inch.

9. The Thumb Tacks. Tacks costing from five to twenty cents a dozen are sufficiently good. Small copper tacks (not iron ones) are sometimes used. The shank of a thumb tack should be slender, the point sharp, and the head shallow and sloping to a sharp edge, that it may not catch the edge of the T-square blade. The shank should be so fastened to the head that, when pressed into the drawing board, it will not push through the head of the tack into the thumb.

10. The Pencils. The common writing pencil is wholly unfit for mechanical drawing ; only the best hexagonal drawing pencils should be used.

11. The Pencil Sharpener (Plate 2). A small, fine-cut file is often used for sharpening pencil and compass leads. A convenient and inexpensive substitute is the sandpaper pad (Plate 2). A piece of fine sandpaper, or emery cloth fastened to a strip of wood, say 6 x $1\frac{1}{2}$ x $\frac{3}{16}$ in., will do very well.

12. The India Ink. The best stick ink gives a glistening, jet black line and is satisfactory for brush work, while lines made with cheap inks are likely to look dead, tend to smear easily, and are unsuitable for brush work. A stick, half size, at fifty cents, is the cheapest that should be considered.

13. The Steel Ink Eraser. A surgeon's scalpel is recommended, as the steel is far superior to that in ink erasers usually found at the stationer's. The length should be about $5\frac{1}{2}$ in., and the shape as shown in Plate 2.

CHAPTER II.

14. The following pages deal with the *technique* and *working methods* of mechanical drawing, independent of the principles of the subject, which are covered by the descriptive geometry.

The study of technique is principally concerned with the training of the several sense powers — manual skill, correct observation, speed, judgment, and taste — which must underlie the operations of the expert all-around draftsman.

Working methods are those dependent upon knowledge rather than upon the sense powers, and, especially, upon experience with the needs and necessities of professional practice. The term also serves to distinguish directly effective and practical methods from those which are purely educational.

The methods here presented are not set forth as the only ones of value. Draftsmen often differ in their opinion of modes of procedure, influenced naturally by their individual experience or by the traditions and character of the work of a particular office. A beginner, however, should take pains to carry out instructions literally until, having learned to work according to the methods indicated, he reaches a point where he is competent to judge other methods.

Success in mechanical drawing rests largely on personal attention to many details. The advantage of workmanlike habits should be kept in mind. Avoid lounging on the drawing table, and other lazy habits. Quiet, brisk attention to the work in hand is not only suggestive of one's personal quality, but is also conducive to better results in drawing. Keep the instruments and materials in orderly arrangement on the table, not only for the sake of appearance, but also to avoid loss of time in searching for the thing needed.

The explanations and directions concerning the uses and care of the instruments should receive close attention. It is not sufficient merely to read the text, but this reading should be supplemented by an immediate examination of the instrument or article considered. The names of the instruments and their parts (Plates 1, 2, and 3) should be remembered. In the directions for manipulation, the processes must not be deferred until the regular exercises are begun, *but each must be practiced at once and exactly as described.* A great deal can be learned by experimenting with the instruments and materials, and through practice before beginning to work on a finished drawing.

15. The Care of the Instruments and Materials. Keep the instruments clean, free from moisture, and always in working order. The pens should be sharpened properly, and must not become clogged with ink. The compass joints should work freely without being either loose or stiff. A drop of oil may be used on the regulating screws of the pens and bow instruments, should they not turn easily. Do not use short leads or blunt needle points in the compass and bow pencil. In opening and closing the bows, pinch the legs together to release the

pressure on the regulating nut (*A*, Fig. 6), then turn the nut, and let the leg spring gently back. When not in use, the bow instruments and the blades of the pens should be left open (*A*, Fig. 7), and the instruments slightly oiled, occasionally, with a soft cloth.

Wood is more likely to warp when one side only is exposed to the air; hence, the drawing board should be left standing on edge, that air may circulate about it, or, if left on the drawing table, it should be closely covered. When not in use, the T-square and triangles should be hung up, and away from heating apparatus and sunlight.

Test, from time to time, the working edge and surface of the drawing board, and the edges of the T-square, and triangles (Arts. 4, 5, and 6); if found inaccurate, they should be sent to a cabinet or pattern maker to be trued.

Fig. 6.

Fig. 7.

If the drawing board is planed off to remove dents, it should be planed on both sides, since it is more likely to warp if planed on one side only.

The best of care should be taken of drawing papers and of drawings, both finished and unfinished, all of which should be kept flat in a portfolio or stiff paper cover. The stick of ink should be nicely wrapped with paper, glued to the ink, to prevent breaking.

16. Drawing Papers. The paper best suited for a drawing depends upon the amount and character of the proposed rendering (Art. 34). In selecting a paper, it may be necessary to take into account its surface — whether hard or soft, smooth or rough,— and its interior quality — whether comparatively soft, or hard and of uniform texture. The Whatman papers are especially satisfactory for finished and display drawing. This paper comes only in sheets, of various weights and sizes (see the end of this paragraph), and is finished in three different styles of surface — *hot pressed, cold pressed*, and *rough*. In a line drawing which is to be inked, which will require considerable time to finish, and which is likely to be subjected to frequent erasure or hard usage, the paper should have a very smooth surface and be of uniformly hard texture throughout — qualities found in the "normal" paper. For precise line drawing, in pencil or ink, without excessive erasure or very hard usage, Whatman's hot pressed and linen record papers are suitable; the former stands erasure better than the latter, but the latter has the smoother surface. For working drawings, and drawings to be traced, an inexpensive but fairly hard paper, such as the "duplex," may be used. This paper is sold in sheets, by the yard, or in ten-yard rolls, and in widths of 30, 36, 42, 56, and 62 inches. It comes in two tints, cream and drab, which may be less trying to the eyes than a white paper. When blue prints are to be made directly from the drawing, a hard bond paper (Crane's, for example) is preferable. Wash drawings must be made on Whatman's cold pressed or on a water-color paper. Useful for practice work and cheaper than the preceding are the German papers, the American imitations, and Manila paper. Besides the above-mentioned papers there are many others, for a description of which the reader is referred to the dealer's catalogue.

Standard sizes of Whatman drawing papers : —

Cap, 13 x 17 inches.	Super Royal, 19 x 27 in.
Demy, 15 x 20 in.	Imperial, 22 x 30 in.
Medium, 17 x 22 in.	Atlas, 26 x 34 in.
Royal, 19 x 24 in.	Double Elephant, 27 x 40 in.

Antiquarian, 31 x 53 in.

Whatman's extra heavy, normal, duplex, and bond papers come only in the royal, imperial, and double elephant sizes.

The better side of a drawing paper is indicated by its water mark. Therefore, in cutting up a sheet, keep track of the better side by putting some distinguishing mark on the *opposite* side.

17. Tracing Paper; Tracing Cloth. For temporary drawings, for transfers, and in planning, tracing paper is well-nigh indispensable. Tracings for blue prints, and tracings required to stand handling and ink erasure should be made on tracing cloth.

The dull surface of the cloth is better for pencil or pen. A wash of India ink or color may be used on the dull surface, if sparingly applied, but the cloth will cockle more or less. If a color wash is used, inking should be done on the opposite or glazed side of the cloth. (Colored crayon works better on tracing cloth than a wash, and is growing in favor; the crayon should be evenly applied to the dull surface, and, if preferred, may be worked flat with a stump.) The glazed surface of the cloth is adapted only for inking, although many prefer the dull surface for this work. When the surface does not take the ink satisfactorily, it should be cleansed with soft paper or sprinkled with powdered chalk (scraped from blackboard crayon). Rub the chalk lightly into the surface, and dust off *thoroughly*. As tracing cloth is very susceptible to moisture, which stretches it, the cloth should not be used when permanent accuracy of drawing is required.

18. Pencils. Ruled pencil lines should be legible, easy to erase, and unaccompanied by grooves in the paper. To satisfy these conditions, it is necessary not only to manage the pencil properly, but to see that the grade of the pencil is adapted to the surface and texture of the paper.

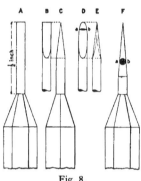

Fig. 8.

In precise drawing — geometrical construction, for example — rendered in very narrow, light lines on a paper hard throughout and having either a smooth or a rough surface, there should be used a 6H pencil sharpened to a ruling point (a, Art. 19). If greater distinctness of line is required — as, for example, when a drawing is to be traced — a 2H to 4H pencil is best. If a surface grooves easily, as is the case with Bristol board and bond paper, which have a hard surface, but are soft beneath, it is best to use an HB or F pencil. For sketching (in mechanical drawing), for suggesting lettering, and for lining in finished pencil drawings, there should be used an H pencil with a somewhat blunt, conical point.

The following list gives a very general idea of the grades of pencil suitable for the papers mentioned in Art. 16 : —

Whatman's Hot Pressed, H to 6H.

Whatman's Cold Pressed, for outline drawing, 6H ; for wash drawing, HB to 3H.

Normal, H to 8H.

Water Color, HB to 3H.

Linen Record, H to 4H.

Duplex, 3H to 6H.

Bond, HB to H.

Bristol Board, HB to H.

19. The Sharpening of Pencils and Compass Leads. (a) *The ruling point.* Cut away the wood of the pencil to expose at least half an inch of the lead (A, Fig. 8). With the file or sandpaper pad resting on something solid, as the edge of the drawing table, and with the forefinger on the wood where it meets the lead (Fig. 9), press the lead lightly but firmly against the sandpaper, and with a steady back-and-forth

Fig. 9.

motion grind opposite sides of the lead to form an accurate wedge (see side view *B*, Fig. 8, and edge view *C*). Next, holding the pencil as in Fig. 10, carry it very

Fig. 10.

lightly back and forth, and with each separate stroke slightly rotate the pencil about its axis in the opposite direction to that of the stroke. This rocking motion forms the finished point (*D* and *E*, Fig 8). It will be noted that, seen sidewise, the contour of the point is elliptical (*D*, Fig. 8) ; a cross section of the point gives the form shown at *ab*, *D*, Fig. 8. The advantage of this ruling point is that, by slightly changing the angle of the pencil, when ruling each new line, the pencil will rest on a perfectly sharp portion of its point.

Another ruling point sometimes used, is formed by first making a conical point (*F*, Fig. 8) and then grinding its opposite sides to form a wedge.

(*b*)   *The measuring point.* This point, which is used to lay off measurements from the scale, is formed by working the lead to a slender and extremely sharp conical point (*F*, Fig. 8). When forming this point, carry the lead back and forth on the sandpaper, meanwhile constantly rotating the pencil about its axis.

It is convenient to have the opposite ends of the same pencil sharpened for ruling and measuring points, in which case the letter representing the grade of the pencil should be scratched or cut at the middle of the pencil.

(*c*)   *The sketching point.* This point is used for putting in an occasional freehand line on the mechanical drawing, for sketching in lettering, writing on the drawing, etc. It should be conical in form, less slender than the point *F*, Fig. 8, and only fairly sharp. For this purpose an H or HH pencil may be used.

Fig. 11.

(*d*)   *Compass leads.* The leads for the compass and bow compass should be placed in the sockets of these instruments and then sharpened. Let the lead extend well beyond the socket of the instrument, that the latter may not come in contact with the sandpaper. Form the point according to the directions for forming a ruling point (see *a*). When sharpened, the lead should be adjusted in the socket thus: Place the pen in the compass, set the needle point of the compass to correspond with the point of the pen, and then set the lead to correspond with the needle point. When the compass is closed (Fig. 11), the needle should project slightly beyond the tip of the lead or pen (see the distance in Fig. 11). When the compass is open (Fig. 12), the plane *ab* of the lead should be perpendicular to a plane passing through the axes *cd* and *ef* of the legs.

20. The Pricker; its Use in Duplicating Drawings. The pricker is used to define line intersections, to lay off measurements from the scale, and in duplicating drawings. A fine needle should be used, and immediately replaced when the point becomes at all blunted.

(*a*) In using the pricker, hold it perpendicular to the paper (Fig. 13). When defining a point in the drawing, do not force the point of the needle through the paper, but make an indentation which is barely visible. In order that such a point may be readily found again, enclose it in a small freehand circle.

(*b*) Diagrams, maps, photographs, etc., are sometimes duplicated as follows: Place the original over one or more fresh sheets of paper, according to the number of duplicates required. Prick the essential points in the original through to the sheets beneath it, and connect the points as in the original. Take special care to hold the pricker perpen-

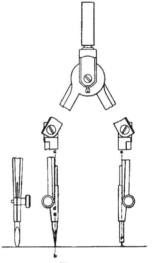

Fig. 12.

dicular to the paper. As a safeguard against connecting wrong points, before connecting the points draw a freehand circle about each point, and letter or number a few of the more important ones.

21. The T-square; Straight Line Ruling. The T-square is used in ruling horizontal lines, and in combination with the triangles (Art. 23).

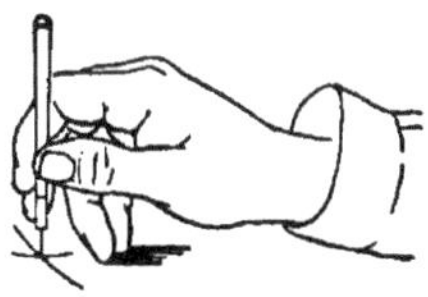

(*a*) *To rule a horizontal line.* Hold the T-square as shown in Fig. 14, press its head firmly against the left-hand cleat of the drawing board, and the blade flat against the paper. Incline the pencil to the right (Fig. 15), and slightly away from the T-square blade (Fig. 16), so that the surface of the

Fig. 13.

ruling point (*D*, Fig. 8) will bear against the blade, and the edge of the point lie in the line of contact of the edge of the T-square blade and the paper. Let the fingers rest on the T-square blade (Fig. 15), press the pencil firmly but lightly on the paper, and carry it steadily from left to right, keeping the pressure uniform, and do not change the initial position of the pencil.

22. The Straight-edge. This is a special ruler, similar to the T-square blade, and used most, perhaps, in surveying and engineering. It is of wood, hard rubber, or steel, and made in various lengths from 12 inches to 120 inches.

(*a*) *To draw accurately a line longer than the straight-edge,* or any especially long line: Stretch taut a fine silk thread between the points which mark the ends of the required line ; prick off carefully one or more intermediate points in the thread, and connect the points by means of a steel straight-edge.

Fig. 14.

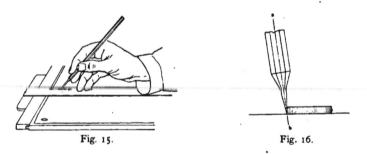

Fig. 15.   Fig. 16.

23. The Triangles are used as rulers, to obtain parallel lines, and to draw lines making certain angles, as shown in the following pages.

TRIANGLE AND T-SQUARE COMBINATIONS.   (*a*)   *To rule a vertical line.* Place the T-square as in drawing a horizontal line.   With the left hand, hold the head of the
T - square   firmly
against the edge
of the drawing
board, and then
slide the hand
from the head of
the T-square along
the blade to keep
it in position.
Place the triangle
with the right hand
(Fig. 17), and hold
it in position with
two fingers of the
left hand — which
also steadies the
T-square blade —
that the right hand

Fig. 17.

shall be free to manage the pencil.   Guide the pencil *away* from the T-square rather than toward it.

(*b*)   *To draw lines making angles of* 15°, 30°, 45°, 60°, *and* 75° *with the*

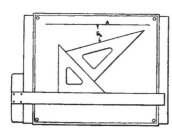

Fig. 18.

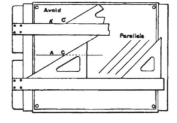

Fig. 19.

horizontal.   Place the T-square and triangles as shown in Figs. 18, 19, 20, and 21.

To draw a perpendicular to a line which makes an angle of 45°, 30°, or 60°, reverse the triangle, as shown in Fig. 20. It will be seen (Fig. 21) that the tri-

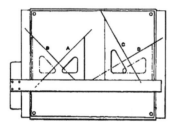 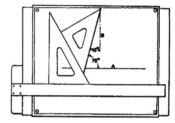

Fig. 20.   Fig. 21.

angle which makes an angle of 75° with the horizontal A must make an angle of 15° with the vertical B.

In drawing a line from a given point in a line, as point C in line AC (Fig. 19), do not fit the T-square to the line, as at $A'C'$, and then attempt to place the corner of the triangle at the given point, as shown at C', but keep the T-square blade away from the given line, AC, that the edge of the triangle shall pass *through* the given point.

TRIANGLE COMBINATIONS. (*c*) *To draw lines parallel or perpendicular to an oblique line, or making with it angles of* 15°, 30°, 45°, 60°, *and* 75°.

Let it be required to draw through point C (Fig. 22) a line parallel to a

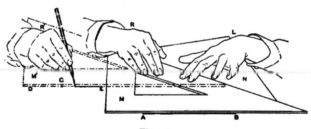

Fig. 22.

given line, as AB. Fit accurately to the given line an edge of either triangle, as M, and place against it another triangle, as N. Hold N firmly in place with the

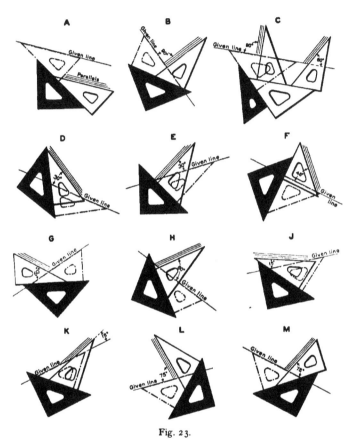

Fig. 23.

left hand, *L*, and slide the triangle *M* along *N* to position *M'*, with its edge passing through the given point *C*. Keep *M'* and *N* firmly in position with the left hand, to free the right hand for drawing the required line, as from *E* through *C* to *D*.

Positions of the triangles for drawing a perpendicular to an oblique line, and for drawing lines making with an oblique line any of the above-stated angles, are shown in Fig. 23. The solid black represents a fixed triangle, and the dash-and-dot, a movable triangle set upon the given line. The full line — representing the triangle which guides the pencil — shows either a new position of the movable triangle (*A*, *B*, *D*, *E*, *F*, *G*, and *H*), or the position of a triangle substituted for it (*J*, *K*, *L*, and *M*). The method *C*, Fig. 23, which uses two movable triangles, is occasionally more convenient, although less direct than the other methods that give the same results.

24. The Protractor. This instrument, used for laying off angles, is a semicircular disc of metal, horn, or cardboard, divided into degrees, half and quarter degrees. Metal protractors with a vernier attachment read to one minute.

25. The French Curve; the Template; Curved Line Ruling. Curved rulers are made of hard rubber, wood, or celluloid, and include the French curve (Plate

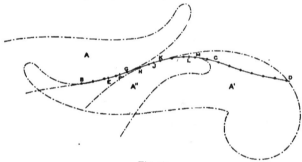

Fig. 24.

2), the railroad curve, the ship curve, and the spline. There is also a patent adjustable curve made of rubber combined with soft lead. The spline and adjustable curves may be bent to fit fairly flat curves up to 60 inches or more in length.

(*a*) *The use of the French curve.* Having located the points in a curve, connect them by a freehand line sketched lightly with a 3H or 4H pencil. This line should pass accurately through the located points and at the same time present a curve as graceful as possible. The line should be sketched without excessive erasure, because this tends to obscure or wholly to obliterate the located points unless they have been defined with the pricker (*a*, Art. 20). The final line, whether drawn in pencil or in ink, should be ruled by means of one or more French curves, as follows : —

Let *BCD* (Fig. 24) be a series of points connected freehand. Find by trial the portion of a French curve which will coincide with or fit the sketched line, and for as long a distance as possible. Rule the line; find a curve which will fit another succeeding portion of the sketched line, rule again, and continue the process until the whole line is thus ruled in sections. In order to insure smoothness of curvature in the line as a whole, in ruling each succeeding section the French curve must be fitted back some distance on the preceding (inked) section. For the same reason, in ruling each section, the line should stop a little short of the full length of the section. These requirements are illustrated in Fig. 24. Curve *A* fits section *BG* of the sketched curve *BCD*, but the ink line is carried only from *B* to *F*. Curve *A'* fits section *GK*, and also fits back on section *BG* to include portion *EF* already inked. The ink line is carried from *F* to *J*; that is, distance *JK* short of the full length of section *GK*. Curve *A''* fits back on section *GK* to *H*, which includes portion *HJ* already ruled, and forward to point *M*. The inked line is carried from *J* to *L*.

In ruling a curve, the pen should be held perpendicular to the paper (Fig. 25), so that, when carried around sharply curving portions of the French curve, it may turn easily on its point.

(*b*) *The template.* This is a substitute for the French curve — made by the draftsman — and used either as a time-saving device or when the French curve cannot be fitted to the given points. It is made of thin sheet metal, wood, rubber, or thick celluloid. If thin wood is used, the required curve may be drawn directly on it. In the case of rubber or celluloid, the curve should be traced and then transferred to the rubber by pricking through the points. The template should be roughed out with a knife or hand fretsaw, shaped more accurately with a coarse file, and finished with a fine file.

Fig. 25.

26. Scales. If the measurements of a drawing are the same as those of the object represented, it is said to be "full size." A drawing is said to be drawn to a scale when, for convenience or from necessity, it is made smaller or larger than the thing represented. For example, a map of the United States may be drawn to a scale of 1 inch = 400 miles; surveyors' plans, to scales of 50 feet to the inch, 80 feet to the inch, and so forth. Drawings of objects are commonly made half size; quarter size, or 3 inches to the foot (written 3 in. = 1 ft.); eighth size, or $1\frac{1}{2}$ inches to the foot ($1\frac{1}{2}$ in. = 1 ft.), etc. Of the various drawing scales manufactured, it is necessary to describe but two, the 12-inch *architect's scale*, divided as stated in the list of materials (Art. 2), and the 12-inch *engineer's scale*, divided to 10, 20, 30, 40, 50, and 60 feet to the inch.

(a) *The architect's scale.* One face of this scale is divided into inches and sixteenths of an inch, like the common foot rule, and is used in making full-size draw-

Fig. 26.

ings. On each side of the other five faces there are two
scales, one being one-half of the other. To illustrate
the arrangement and reading of the several scales,
consider the scale of 1 in. = 1 ft. (Figs. 26 and 27).
The denomination of the scale is indicated by the
numeral " 1 " placed at its right-hand end. The scale
is divided into 12 equal parts, each representing 1
foot, and the last one, *CD*, Fig. 27, is divided into 24 equal parts, representing inches
and half inches; that is, this 1 inch, representing 1 foot, is divided proportionally

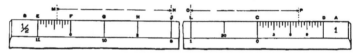

Fig. 27.

to a foot rule, except that the smaller divisions are omitted. Beginning at the zero
(under *C*) and reading to the right, groups of 3 inches each are indicated by the
numerals " 3," " 6," and " 9." Reading from the zero to the left, feet are indicated
by the numerals, as " 1," " 9," " 10," and " 11," placed on the concave surface immediately below the face of the scale.

To read a measurement. Let it be required to measure the line *OP*, Fig. 27.
The scale must be placed so that the feet and inches can be read continuously without moving the scale. On applying the scale to the given line, it will be quickly
seen that the line contains feet and inches; in this case, the graduation line " 1 "
(under *L*) must fall on the point *O*; the length of *OP* is 1 ft. $7\frac{1}{2}$ in. In the
scale $\frac{1}{2}$ in. = 1 ft. (Fig. 27), it will be seen that the smallest divisions of *EF* (= $\frac{1}{2}$
inch) represent inches, and that these read from the zero (under *F*) to the left.
The feet are read from the zero to the right. The line *MN*, read from this scale,
measures 3 ft. 5 in.

(b) *The engineer's scale.* The face (Fig. 28) selected to illustrate the several

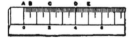

Fig. 28.

graduations of this scale is divided into 600 equal parts. Each inch is therefore
divided into 50 equal parts, each representing 1 foot, and the scale is said to be 50

feet to an inch, indicated by the "50" (under F) stamped on the scale. It will be seen that groups of five divisions, as AB (Fig. 28), and of ten divisions, as DE, are denoted by the greater length of the graduation lines, and that each group of twenty divisions is indicated by a numeral stamped on the scale — as the "2" opposite C, which is read 20 feet. It will also be seen that inches are not distinguished by number ; if required, these must be found by means of the group divisions and numbers. Thus, for example, reading from the left-hand end of the scale, 1 in. is obtained by taking the graduation line, E, midway between the "4" and the "6."

To find readily the face of the scale in use, a guard (see Fig. 26) is convenient. *The scale should never be used as a ruler.*

27. The Compass.— Large circles and circular arcs are drawn, according to length of radius, with the compass, the compass with the extension bar, or the beam compass.

For small circles, of ¾ inch radius or less, the bow compass should be used.

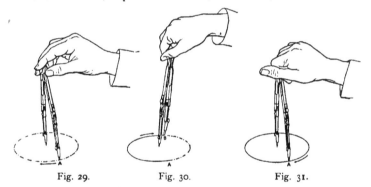

Fig. 29.   Fig. 30.   Fig. 31.

Use the shoulder tip of the needle point (Plate 1), and have the lead of the compass sharpened and adjusted in the socket according to d, Art. 19. For drawing very small circles, the inside of the needle point should be beveled.

(*a*) *To describe a circle.* Set the compass to an approximate radius, with the legs bent to bring each perpendicular to the paper (Fig. 12), and then set it to the exact radius. Holding the head of the compass with the tips of the thumb and first two fingers (Fig. 29), with the needle point merely resting on the paper, start the circle at point A, under the inner edge of the wrist, and describe a circle with one continuous sweep of the lead. In describing a circle, incline the head of the compass slightly forward in the direction of its motion — indicated by the arrows, **Figs. 29, 30, and 31,**— and guide the instrument by a combined finger, wrist, and arm movement, during which the head of the instrument should roll between the thumb and forefinger (Figs. 29–31). Do not acquire the habit of carrying the lead back and forth over a circle; once drawn, let the circle stand — improvement should come from practice, and not from going over or patching a line. If a darker

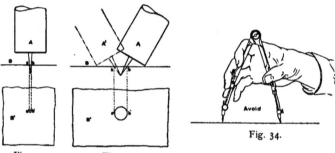

Fig. 34.

Fig. 32.   Fig. 33.

line is required, change to a softer lead. Furthermore, do not thrust the needle point into the paper, but give it only sufficient pressure to keep it from slipping.

The need of bending the needle-point leg and of inclining the compass but slightly in describing a circle is shown in Figs. 32 and 33. When the needle point is perpendicular to the paper (A, Fig. 32), if the paper is accidentally punctured, the center thus made will have a diameter, $a'b'$, equal only to the diameter, ab, of the needle point. If, however, the needle point is inclined, as in A, Fig. 33, the

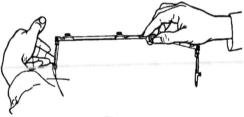

Fig. 35.

swinging of the compass will cause the needle point to ream out the center, and thus prohibit its further use for accurate work.

Never use the compass as shown in Fig. 34.

(*b*) *The lengthening bar.* When the lengthening bar (Fig. 35) is used, the compass is likely to spring under pressure; hence it must be handled lightly. Bend the legs to bring them perpendicular to the paper, incline the instrument slightly in the direction of its motion, and at the beginning and the end of the line guide the describing leg with a finger of the left hand.

28. The Beam Compass. This instrument (Plate 3) is practically a compass with a separated head (*A* and *B*, Plate 3) which slides on a wooden· bar. Slight corrections of distance between the legs of the compass may be made by means of a spring which is regulated by the nut *C*.

29. The Hair-spring Dividers; the Bow Spacers. The hair-spring dividers and the bow spacers (Plate 1) are used to transfer distances, and to space or divide lines into equal parts. The points of the instruments should be very sharp, of exactly the same length, and, when closed, should come together accurately. If the points are blunted or of unequal length, the defect should be remedied on the oil stone.

(*a*) *Spacing.* Before commencing to space, ascertain the range (*BC*, Fig. 36) of the hair-spring leg by turning the screw *E*, and then set the point of the leg at *A* midway between *B* and *C*, so that the leg may be moved by means of the hair spring in either direction according to need. Let it be required, for example, to divide line *D'B'* (Fig. 37) into 13 equal parts. Moving the legs by means of the head joint, set the dividers approximately equal to $\frac{1}{13}$ *D'B'*; then, starting with one leg of the dividers placed at *D'*, step along the given line until the 13 spaces are laid off. If, on laying off the last, or 13th, space, the leg of the dividers extends beyond the given line, as distance *B'A^{vii}*, the assumed distance in the dividers must be decreased by $\frac{1}{13}$ of the excess, *B'A^{vii}*, determined by judgment. If, however, on laying off the 13th space, the leg of the dividers does not reach the end of the line, as at *C*, it is evident that the distance

Fig. 36.

in the dividers must be increased by $\frac{1}{13}$ of the deficiency *CB'*. In making either correction, if the error is a considerable one, the legs of the instrument should be moved by means of the head-

Fig. 37.

joint, but for a slight error use the hair spring, which, as already stated, should be set before beginning to space, so that the leg may at once be moved in either direction. Until a close approximation to a required spacing unit is obtained, instead of stepping the dividers along the given line, it is best to keep the points a little to one side of the line in order not to mar it. In stepping off the spaces, care should be taken not to spring the dividers, and thus change the distance between the points of the instrument — a result which will follow if the points are pushed into the paper. When a close approximation to the required spacing unit has been obtained, the trial spacing should be transferred to the given line; but the paper must not be punctured or perceptibly indented, since the points of

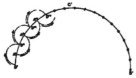

Fig. 38.

the dividers are very likely to slip back into the holes, and thus to frustrate all efforts for a satisfactory result. To minimize error from failure to keep the points of the dividers exactly on the given line, let the advancing leg swing in alternate directions, as indicated by the lettering and arrows (Figs. 37 and 38). When indicating a final spacing, the paper should be merely indented, not directly, but by going over the given line several times, each time with just enough pressure to indent the paper visibly after the several repetitions.

When stepping with the dividers, keep the plane of its legs perpendicular to the paper.

(*b*) *Setting the dividers (or compass) for a scale measurement.* When setting the dividers for a scale measurement, hold the scale in the left hand, with the face,

as *ADC*, Fig. 39, horizontal, meanwhile holding the dividers so that the plane of its legs will be perpendicular to the face of the scale. Or, place the scale on the drawing, with the wrists resting on the scale, to keep it in place, and to free both hands to manage the dividers. *Never hold, with both hands, the compass or dividers between the scale and the body* (see the lower side of Fig. 39), as the position is an awkward one, the numbers stamped on the scale are inverted, and the scale will slip. In taking a measurement

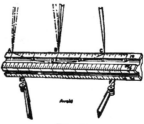

Fig. 39.

from the scale with the dividers, slight errors may readily be detected by stepping the dividers along the scale for a considerable distance, and then noting whether the total distance stepped off is an exact multiple of the required distance.

30. Proportional Dividers. This instrument (Plate 3) is occasionally very convenient in making reduced or enlarged copies of drawings, photographs, etc. A required ratio, as between distances *ab* and *cd* (Plate 3), is obtained by means of the sliding joint *A'*.

31. India Ink. This medium — in line drawing applied with the ruling and compass pens — is used to give finish to a pencil drawing, and for greater distinctness and durability. It is also used in wash drawing, as described in Chapter X. The stick India ink gives the best results. Writing inks are wholly unsuitable for drawing. The ready prepared drawing inks are not here recommended for finished drawing; they do not flow so well as the freshly ground ink, are not always strictly black, and are likely to injure the paper sufficiently to cause fuzzy lines.

(a) *To prepare India ink for line drawing.* Fill the ink slab (Plate 2) with water just sufficient to overflow the ink well. Hold the stick of ink vertical, and with a moderate pressure give it a rotary motion around the edge of the ink well. Continue grinding until the ink is *absolutely black*, but not thick; this will require perhaps five to ten minutes, according to the hardness of the ink and the amount of water used. As the grinding proceeds, test the condition of the ink by ruling lines on paper of the kind upon which the drawing is to be made. Do not judge the ink before it is thoroughly dry. The stick of ink should be wrapped as directed in Art. 15, and wiped perfectly dry after use; otherwise, it is likely to crumble. Keep the liquid ink covered to prevent thickening and to keep out dust. *Do not set the ink slab on the drawing or the board, but keep it well away from the work.*

32. The Ruling Pen. (a) *Manipulation of the pen.* To fill the pen, let it rest in the ink a moment, then carefully wipe off the outside of the blades with a rag or the chamois leather. A better way, is to fill between the blades with a common writing pen or quill. Always, before beginning to ink, and on changing the width of a line, the pen should be tried on the edge of the sheet, to see that the ink flows freely and that the line is of a desired width. Avoid the habit of touching the pen to the lips, to make the ink flow; if two or three trials of the pen on the paper, the finger, or the drawing board — where the pen should follow the grain of the wood — will not cause the ink to flow, either the ink is too thick or the pen needs cleaning.

When using the pen, hold it perpendicular to the paper (Fig. 40), or nearly so, with the tip of the forefinger resting on the head of the regulating screw. If the pen is inclined in the direction of its motion, it is difficult to keep a line uniform in width, or to end a line accurately at a given point. Carry the pen from left to right steadily, rather slowly, and without the slightest rotation about its axis. To satisfy this last condition, the position of the hand

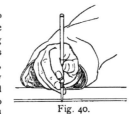

Fig. 40.

relative to the T-square or to the ruler must not change; there must be an arm movement from shoulder and elbow, but no wrist motion. In drawing a long line a slight movement of the wrist appears to be necessary, but this should not change the position of the hand relatively to the ruler. If the handle of the pen is inclined outward, in order to bring the point into the line of contact of the ruler and paper, the ruler, especially if made of rubber or steel, is likely to attract the ink and thus blot the paper; hence the blade should be kept slightly away from the lower edge of the ruler, unless the latter is provided with a beveled edge. Blotting paper should always be at hand in case of accident.

When filling the pen, do not hold it over the drawing.

(b) *The care of the ruling pen.* As the blades of a good pen are highly tempered and therefore brittle, care must be taken, when filling the pen, not to strike the points against the ink well; also, in setting the pen, not to screw the blades together too tightly. If a pen works badly, the blades should be slightly separated and examined with a magnifying glass to see whether the pen is dull, the blades of unequal length, or their points broken off. A blunted point may also be detected by viewing the blades endwise; if it is dull, a light spot, due to reflected light, will be seen at the tip of the point. Every student should learn to sharpen his own pen, since inability to do this may prove to be a serious handicap if he chances in the future to be located at a point remote from an instrument repair shop.

(c) *To sharpen the ruling pen.* Use a fine oil stone and plenty of oil. With the blades closed, hold the pen perpendicular to the face of the oil stone,

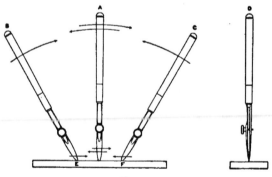

Fig. 41.

as indicated by views *A* and *D*, Fig. 41. Keeping the pen in a plane perpendicular to the surface and parallel to the long edges of the oil stone, carry the pen back and forth, and at the same time rock the pen, as indicated by positions B and C, Fig. 41. This grinding should bring the blades to the same length, and so shape the point of the pen that, when seen as at A, Fig. 41, the outline of the point will appear elliptical. Now grind separately the point of each blade as follows :

Hold the pen as shown at A, Fig. 42, making an angle of about 10° or 12° with the surface of the stone. Carry the pen back and forth as indicated by the arrows, and, while so doing, rotate it about the axis of its handle. In this rotation the right-hand edge of the blade, in passing from A to A', gradually approaches the surface of the stone. At A' the pen is rotated to a position corresponding to the initial position (see B), and in the stroke from B to B' the left-hand edge of the blade gradually approaches the stone. The pen is again rotated to the initial position A, and the motions are continued. The curved portion of a cross section of the point should be approximately elliptical. As the grinding proceeds, the point should be examined

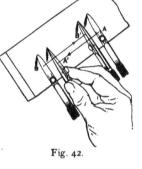

Fig. 42.

from time to time with the reading glass (Plate 2), and the pen tested by lines carefully ruled on drawing paper. When properly ground, the blades of a pen have exactly the same length and sharpness; if one blade is longer, or sharper than the other it will be difficult to guide the pen, as it will persist in taking a course of its own. As a rule, care, patience, and considerable time are necessary in order to grind a pen properly. *Never grind the inside of the blades.*

(*d*) The point of the small pen (Plate 1) should be as sharp as possible without actually cutting the paper, and this pen should be used only for fine lines. The larger pen, with its greater ink capacity, is better adapted for general work; the point should be only moderately sharp.

33. Erasure. (*a*) *Pencil line.* Use rubber which is clean and pliable. When erasing on thin paper or tracing cloth, special care must be taken not to crumple it; hold down firmly with the thumb and finger, and erase between. For erasure within a small space, shave the rubber to a point or wedge, and use a card to protect adjacent lines of the drawing. Sponge rubber is convenient for large areas.

(*b*) *To remove ink lines from drawing paper.* In office practice, instead of a new drawing, changes in design are frequently made on old drawings by means of erasure and redrawing. In such work, and in ordinary corrections, removal of blots, and drawing over an erasure, the work should be so nicely done that it will in nowise deface the drawing — a matter requiring knowledge, care, and skill. Every vestige of ink should be removed. Both the steel eraser and the sand rubber (rubber ink eraser) may be used to advantage, and the first step toward skilful erasure lies in keeping the steel eraser very sharp. The scalpel (Plate 2), if a new one, should be reground, and afterwards kept to a keen edge by frequent whetting on the oil stone. The form shown in Fig. 43 is recommended. This edge, but slightly convex, is best adapted for erasing areas ; while the point, if kept sharp, is very satisfactory for minor erasures and cleaning up ragged lines.

Before beginning to erase a line, note with the finger whether there is a considerable deposit of ink. If so, first use the steel eraser to remove as much of the ink as possible, but without scraping the paper, and then apply the sand rubber. In using any eraser, the principal thing is not to make grooves or ruts in the paper : hence the eraser should be carried in the direction of the line to be erased, not crosswise, and over the paper for some distance on all sides of the line, so that the paper may be worn away uniformly. To the same end, in converting a full line into dashes, carry the steel eraser back and forth in the direction of the line, *not crosswise.* To prevent ink from spread-

Fig 43.

ing, when drawing over an erasure, the surface of the paper should be thoroughly but lightly polished with the burnisher (Plate 2). If the burnisher is applied too strongly, the drawing board will yield, and cause a rut in the paper. To obtain the best result, place the paper on a hard surface, such as the rubber triangle or a piece of glass. Excessive gloss due to burnishing may be removed with the soft rubber ; but, if lines are to be redrawn, the gloss should not be removed until the lines have been put in.

(*c*) *The erasing shield.* This device is paper, cardboard, or celluloid, in which are cut slits and holes of various shapes and sizes. The shield is placed over the drawing, so that only the portion to be erased will be exposed to the rubber. While shields may be procured of dealers in drawing supplies, it is usually more convenient to keep a sheet of thin celluloid on hand, and to cut the apertures according to immediate needs.

The use of the device is illustrated in Figs. 44, 45, 46, and 47. Fig. 44 shows a series of blots and mistakes at *E, F, . . . M*, supposed to have been made in a drawing of the locomotive hand-rail stud, Plate 12. In Fig. 45 are given the forms of the edges and of the apertures in the erasing shields appropriate in this case. Fig. 46 represents the drawing after erasure, and Fig. 47 the drawing with the lines restored. The edge *M'* (Fig. 45) is used in reducing the excess width of the lines M (Fig. 44). The full line lettered E, Figs. 44, 45, and 46, is a mistake in inking, since, as this line represents an invisible edge of the object, it should be composed of dashes (E, Fig. 47). The line lettered H (Figs. 44, 45,

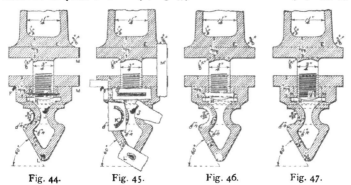

Fig. 44.　　Fig. 45.　　Fig. 46.　　Fig. 47.

and 46) represents a visible edge of the object, and therefore should be made a full line as at H (Fig. 47).

(*d*) *To remove ink lines from tracing cloth.* Slip cardboard, glass, or a triangle under the cloth, for a backing, and then remove accumulations of ink with the steel eraser, which must not come in contact with the cloth. Complete the erasing with a ruby or emerald rubber, using an erasing shield if necessary. Before redrawing, apply talc ("metal worker's crayon"), and polish with the burnisher. Erasure with a soft rubber is easier on the glazed than on the dull surface of the cloth; but in no case should the sand rubber be used, as it will quickly make a hole in the cloth.

CHAPTER III.

34. Rendering. This term signifies the material qualities of a drawing, and the manner of its execution. The material qualities include the drawing surface — as wood, metal, or paper,— and the medium employed — as chalk, graver, pencil, or ink. The manner of execution may rest wholly upon stated, definite methods — as in the manipulation of the instruments, Chapter II.— or, besides this, may call for individual ingenuity, judgment, or taste.

35. Precision and Speed. The degree of accuracy with which a drawing should be rendered depends upon the purpose it is to serve. For example, the plotting of surveys and the graphical solution of engineering problems often require the greatest possible accuracy, to secure which it may be necessary to use a high-power reading glass. Working sketches and architectural sketch rendering are examples of the opposite extreme, or free treatment ; here the principal consideration is speed. The greater part of the practical requirement in drawing, however, lies between these two extremes, and includes both accuracy and speed.

At the outset, whatever the character of his prospective work in drawing, the student should learn to draw *accurately*. As he progresses, he should learn to adapt himself to varied requirements, and should acquire the judgment necessary to work to the best advantage.

Generally speaking, accuracy of rendering implies diminished speed. It does not follow, however, that in order to obtain accuracy it is necessary to work slowly. From the very start there should be a sustained effort to work accurately and rapidly ; but fussiness must not be mistaken for the care necessary to secure accuracy, nor mere bustle for speed. Speed depends upon quick judgment and right methods rather than upon quickness of hand.

36. Refinements in Observation and Manipulation. The following methods properly supplement the methods considered in Chapter II., when extreme accuracy is required.

(*a*) *The line of sight* is an imaginary straight line drawn from the eye to any observed point. It is important that the line of sight shall be perpendicular to each observed portion of the drawing ; thus, in viewing the drawing paper in the vicinity of *B*, Fig. 48, the line of sight, *AB*, should be perpendicular to the paper. This condition will be met best if the student stands, when drawing, so that he may easily move about as necessary to bring the line of sight to the position indicated. A block to incline the drawing board is convenient.

(*b*) *When ruling lines*, the draftsman's head should move with the ruling point. If the head remains in a fixed position and only the eye follows the ruling point, it is evident that, as the point

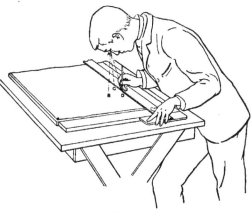

Fig. 48.

moves away from the eye, the angle which the line of sight makes with the paper constantly decreases, and, consequently, the chance of error increases. Also, in ruling, the eye should see the line of contact of the ruler and paper; hence the line of sight, *AD*, cannot be exactly perpendicular to the paper, but the angle *BAD* (Fig. 48) should be as small as possible.

In the case of a long line, the ruling should be done from a standing position (Fig. 48). The draftsman should stand facing, and with the left side of the body turned slightly toward, the drawing. Leaning over the drawing, the line of sight, *AC*, should be directed slightly forward and inward from the perpendicular *AB* — forward, in order to see in advance of the moving point, *D*, and inward, to see distinctly the line of contact of the ruler and the paper. Standing as described, in drawing a line the draftsman may walk *forward*, whereas, if he stands with the right-hand side of the body turned toward the work, he is obliged to walk *backward*.

(*c*) *In laying off a measurement from the scale*, the line of sight must be as nearly perpendicular as possible to the edge of the scale at each of the two graduations which determine the measurement; that is, in pricking off a graduation, as *C*, Fig. 49, the angle *BAC* (or *BAD*) between the perpendicular, *AB*, to the edge of the scale, and the line of sight, *AC*, should be no greater than is necessary to bring into view the point of the pricker, or pencil.

(*d*) *Precision in noting line intersections.* In precise rendering it is not sufficient, when noting an intersection, merely to glance at the point, but it must be

closely scrutinized in order that the point of the pricker may be placed at the intersection of the imaginary center lines of the lines of the drawing. To illustrate, let the black lines, *B*, Figs. 50 and 51, represent an enlargement of the lines *ab* and *cd*, *A*, Figs. 50 and 51, and the white lines (*B*) represent the imaginary center lines of lines *ab* and *cd*. The intersection of the white lines represents the exact point at which the point of the pricker must be placed when noting the intersection, *e*, of the lines *ab* and *cd*.

Fig. 49.

In defining a tangent point, as *g*, Fig. 52, the position of the point of the pricker should correspond to the intersection of the vertical and horizontal white lines in *B*.

(*e*) *Precision in laying off scale measurements.* The point of the pricker

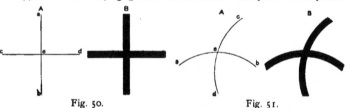

Fig. 50.   Fig. 51.

must coincide exactly with the intersection of the center line of the graduation and the edge of the scale. (See *B*, Fig. 53, where the black areas *a'*, *b'*, . . .*f'* represent enlargements of the graduations *a*, *b*, . . . *f* (*A*).)

(*f*) *Precision in spacing with the dividers.* It is not sufficient that the points of the dividers fall somewhere near a given line, or even somewhere on this line,

Fig. 52.   Fig. 53.

but they must be placed exactly on the imaginary center line of the given line, using a reading glass (Plate 2), if necessary. To illustrate the importance of this, let it be supposed that the points of the dividers fall on opposite sides of the given line, as at the intersection of the white lines with the edges of the black area in *C* and *F*, Fig. 54. It is evident that the distance in the dividers will be laid off in a zigzag line, and that, if the points of the dividers fall several times on one edge of the given line, then several times on its opposite edge, next on its imag-inary center line, and so forth, a

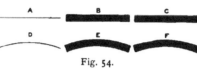

Fig. 54.

very slight error will occur each time the dividers is advanced. This local error, however trifling, will be multiplied in stepping the required distance, and may result in a perceptible accumulated error.

(*g*) *Precision in drawing circles*. A compass setting taken from the scale must be tested by drawing a trial circle and scaling its diameter.

In the case of a large number of equal circles the diameter must be carefully tested from time to time, with the scale, to see if the resharpening of the compass lead has changed the radius.

37. Conventions. (*a*) *Line conventions*. Line conventions commonly used in mechanical drawing are given in Fig. 55. The cut shows the actual width of line,

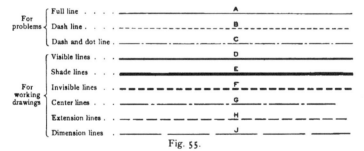

Fig. 55.

length of dash, and space between dashes satisfactory for general work. The student should become familiar with the several differences in width of line, length of dash, and space between dashes, by a careful comparison of the conventions. More marked distinctions may be obtained by using colored inks, a method which also saves the time of making dashes, as, when color is used, all lines are usually made full.

The conventions shown in Plate 4 are used more or less in practice, and, as an exercise in rendering, afford excellent training.

(*b*) *Breaks*. When only a portion of an object is represented, this fact may be indicated by an irregular line, as, for example, at *F*, Fig. *A* (Plate 4), and the drawing is said to be "broken off." To economize space, an object may be represented as broken and brought together, as in the first and last examples, Fig. *A*. A break in a metal piece rectangular in section may be indicated according to *E* and *E'*; a break in a wooden piece, according to either of the two styles of treatment *H* and *H'*. A cylindrical piece may be broken as shown in Fig. *B*, and the surface of the break cross hatched or not, according to personal preference or the custom of the office.

(*c*) *Graining*. A wooden piece seen lengthwise may be grained as shown at *F*, Fig. *A*, an end view as shown at *F'*.

(*d*) *Section lining or cross hatching*. The conventions, Fig. *C*, indicate material as follows : —

J, Cast Iron.	*L*, Wrought Iron.	*N*, Composition.
K, Cast Steel.	*M*, Wrought Steel.	*O*, Babbitt.

The series *P*, *Q*, . . . *V*, is the same as the series *J*, *K*, . . . *O*, except that the areas are larger, which requires that the lines of the cross hatching shall be coarser and more openly spaced.

(*e*) *Conventional line shading* (*Fig. D*), is used in drawings for cuts of mechanical objects and, to a very limited extent, in engineering drawing (see *c*, Art. 66). The number of lines, and their spacing and respective widths in a piece of shading, depend upon the size of the shaded area and the effect desired.

38. Rendering of the Conventions, Plate 4. (*a*) *Breaks and graining.* These conventions (Figs. *A* and *B*) must be rendered with a coarse writing pen, wholly freehand, and in flexible line. The outline of the break, Fig. *B*, should be of uniform width, curved throughout its entire length, and drawn tangent to the contour elements of the cylinder. In first attempts at rendering these conventions, the plate should be closely copied.

(*b*) *Cross hatching* (*Fig. C*). The lines of the several cross hatchings should be drawn at 45° with the horizontal, except in convention *O*, where the lines are drawn at 60°. Place the proper triangle against the T-square, and — as preparation for general office practice — space the lines *wholly by eye*, or by the sense of touch — that is, by feeling or *sensing* the amount the triangle should be moved to obtain the required spacing.

Conventions *L* (*R*) and *M* (*S*) should be treated in the following manner : Let *R'* represent the section lining at the upper right-hand corner of example *R*. Set the pen for the narrow lines (*R*), and, keeping the same width of line, represent the wider lines (*R*) by double lines, as at *D* and *E* (*R'*). Cross hatch the whole surface, and then fill in the narrower spaces to make the wide lines. Make the

Plate 4
(Study Plate 3)

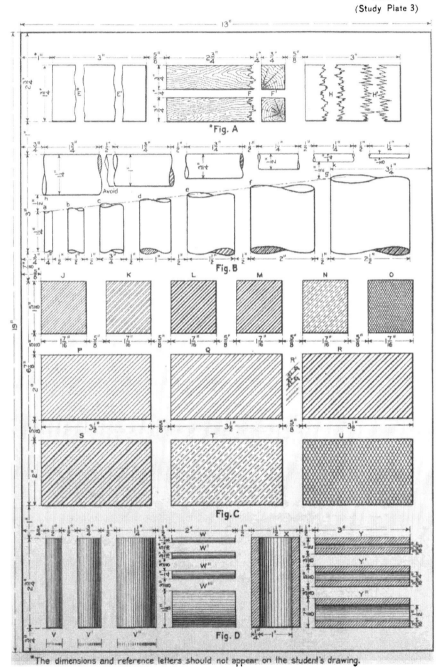

*Fig. A

Fig. B

Fig. C

Fig. D

*The dimensions and reference letters should not appear on the student's drawing.

(41)

spaces, as *A*, *B*, and *C* (*R'*), between the narrow lines and the edges of the wide lines equal in width.

Some draftsmen insist that if a break is cross hatched (see Fig. *B*), it should be done freehand.

The instrument known as a section liner may be convenient, or perhaps necessary, in making fine drawings and cuts; but *this instrument must not be used in the exercises here given.*

(*c*) *Conventional line shading* (*Fig. D*). The spacing and widths of line must be determined wholly by eye. A preliminary penciling of a narrow strip of the shading (as shown below *V*, *V'*, and *V''*, Fig. *D*), to establish the number of lines, spacing, and line widths, should be indicated; but, with this exception, the shading must be done directly with the ruling pen. Begin at the left-hand border of a vertical piece and at the upper border of a horizontal piece.

In the case of a cylinder first set the pen for a very narrow line, and, while changing the spacing, retain the same width of line until the wider lines at the right-hand third of the cylinder are reached. Then, besides changing the spacing, gradually increase the width of line by changing the setting of the pen. The widest lines of the shading may need to be built up by going over each several times, in which case let the lines dry frequently in order to avoid blots. When all of the shading has been thus laid in, the narrower lines may be retouched, if necessary, and the spacing and line widths of the right-hand third of the cylinder corrected.

Excellent examples of line shadings in great variety may be found in catalogues of machinery, etc., illustrated with fine woodcuts. In first attempts, an example should be closely copied, not line by line, but for its general effect.

39. Lettering. In connection with the study of mechanical drawing it is important to have extended practice in lettering and dimensioning (Art. 42).

(*a*) *Styles and sizes of letters.* In Plate 5 are given vertical and inclined Gothic letters and numerals in two sizes and widths of line, which should be used in rendering data and dimensions on the practice drawings that follow. These styles should also be used for titles, but the letters must be larger.

There should be lettered on the margin of each practice drawing the word "*Plate*," *the number of the plate*, the *student's name*, and the *date* (when a drawing is completed). The required heights of the letters, for two sizes of border line, are given in Fig. 56.

The general effect of the lettering and dimensions of a drawing should correspond with the general effect of the drawing. That is, if the drawing is strong in appearance, a bold letter should be used; but, if a drawing is light or weak in its general effect — as in a geometrical construction — a lighter letter should be selected (compare Figs. *A* and *C*, Plate 5).

40. Stroke Letters. (*a*) *Rendering.* In office practice it is commonly required that small titles, reference letters, and dimensions shall be rendered with

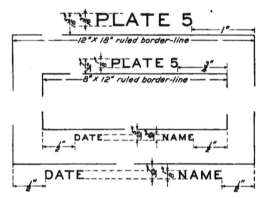

Fig. 56.

direct strokes of the pen, very rapidly, and without the use of guide lines. A system of stroke letters is given in Fig. *J* (Plate 5) : the arrows show the direction in which the pen should be carried ; and the numerals, the order in which the strokes should be made.

The practice drawings here given must be lettered and dimensioned according to this system, unless otherwise specified. To acquire proficiency, the student should practice daily the alphabets, Plate 5, until he can make the letters and numerals with a fair degree of speed and uniformity. For a line of letters or numerals, one base or guide line, only, may be ruled.

(*b*) *Lettering pens.* For rendering dimensions and data — when the letters and numerals have a width of line corresponding to that of the letters in Fig. *C* or *D*, Plate 5 — a No. 303 Gillott's pen is recommended. Letters having a considerable width of line — as wide as that in Figs. *A* and *B* — may be rendered, according to preference, with a coarse writing pen, a ball-point pen, a turned-point pen, or a ruling pen specially ground for lettering.

(*c*) *Letter rendering with the ruling pen.* A reproduction of a piece of rapid stroke lettering with the ruling pen is given in Fig. 57. The cut is half the size of the original, which was taken from a large number of office drawings equally well lettered, and in a like manner.

Plate 5

ABCDEFGHIJKLMNOPQRSTUVWXYZ
1234567890& abcdefghijklmnopqrstuvwxyz

Fig. A.

ABCDEFGHIJKLMNOPQRSTUVWXYZ
1234567890& aabcdefghijklmnopqrstuvwxyz

Fig. B.

ABCDEFGHIJKLMNOPQRSTUVWXYZ
1234567890& abcdefghijklmnopqrstuvwxyz

Fig. C.

ABCDEFGHIJKLMNOPQRSTUVWXYZ
1234567890& aabcdefghijklmnopqrstuvwxyz

Fig. D.

ABCDEFGHIJKLMNOPQRSTUVWXYZ
1234567890& abcdefghijklmnopqrstuvwxyz

Fig. E.

ABCDEFGHIJKLMNOPQRSTUVWXYZ
1234567890& aabcdefghijklmnopqrstuvwxyz

Fig. F.

ABCDEFGHIJKLMNOPQRSTUVWXYZ
1234567890& abcdefghijklmnopqrstuvwxyz

Fig. G.

ABCDEFGHIJKLMNOPQRSTUVWXYZ
1234567890& aabcdefghijklmnopqrstuvwxyz

Fig. H.

ABCDEFGHIJKL
MNOPQRSTUVWX
YZ 1234567890&
abcdefghijklmnop
qrstuvwxyz

Fig. J.

$2\frac{3}{8}"$ $5\frac{1}{8}"$ $6\frac{7}{32}"$ $2\frac{3}{8}"$ $5\frac{1}{8}"$ $6\frac{7}{32}"$ < > $\leftarrow 0'-5\frac{1}{2}" \rightarrow$

Fig. K.

(45)

The special advantage in lettering with the ruling pen is the increase obtain-able in the size of a letter and its width of line. An inexpensive pen, or one which has proved unsatisfactory for ruling, is sufficiently good for the purpose.

(*d*) *To grind a ruling pen for lettering.* Hold the pen perpendicular to the oil stone and grind until the point is very blunt. Next hold the pen as in Fig. 42,

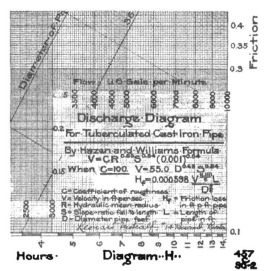

Fig. 57.

and grind until the two blades, taken together, are conical in shape, except at the point, which should remain blunt. To shape the point, start with the pen held perpendicular to the stone, and carry its point in a circular path, meanwhile con-stantly changing the direction of the pen to all angles from the perpendicular to 45° with the stone — a movement which changes the roughly blunted point to one having a spherical shape.

(*e*) *Manipulation ; lettering with the ruling pen.* Hold the pen at any angle between 45° and 60° with the paper, with the regulating screw horizontal and the thumb resting against the head of the screw. Render the letters according to *J*, Plate 5. It is best to fill the pen with the common writing pen, or a quill, as wip-ing is likely to absorb too much ink from between the blades.

41. Drawn Letters. When letters and numerals are gradually built up or developed by stages — as is necessary for the most finished results in freehand lettering — they are said to be *drawn*.

(a) *Rendering of drawn letters.* Guide lines must be ruled as shown in Figs. 58 and 62. All letters and numerals throughout the drawing should be developed

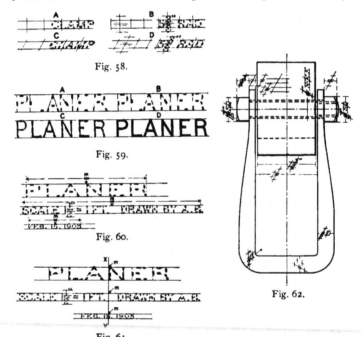

Fig. 58.

Fig. 59.

Fig. 60.

Fig. 61.

Fig. 62.

wholly freehand, in the following order. (I.) Suggest in pencil (*A*, Fig. 59) the letters and numerals — for their sizes and position relatively to the drawing or its parts. (II.) Correct the pencil suggestions, for the spacing, form, and verticality, or slant, of the letters (*B*, Fig. 59). (III.) Without changing the treatment (*A* and *B*), ink in the suggestions throughout the drawing, but at the same time make, if necessary, further corrections in the form and verticality, or slant, of the letters and numerals (no figure). (IV.) Connect or pull together the blocks of the suggestions

(*C*, Fig. 59). (V.) Bring the lines to the required width and complete the letters, making all edges sharp and true (*D*, Fig. 59).

(*b*) *Balancing a title, or a line of letters.* When one or more lines of letters must be balanced on a given line, as *XY*, Fig. 61, proceed as follows: Lay off

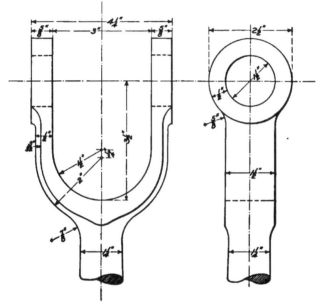

Fig. 63.

accurately the heights of the letters, and rule the guide lines (Fig. 60) on a slip of paper, but without regard to their position. Suggest each line of the title (Fig. 60). Find the middle point, as *m*, of each line of letters; fit the middle point, *m*, on the center line, *XY*, of the final title (Fig. 61); mark off the letter widths, and develop the letters by stages as indicated in paragraph *a*.

42. Dimensioning. This term signifies the giving of measurements on drawings (Fig. 63), and includes the rendering of the numerals, their arrangement on the drawing, and the selection of their size and style.

A dimension line is the broken line connecting the arrow heads which indicate the points of a measurement. *Extension lines* (see the vertical lines composed of short dashes, Fig. 63) are used, in case of interference or confusion, to carry points of measurement to another part of the drawing. Line widths and length of dashes for both dimension and extension lines are given in Fig. 55.

Numerals of a size suitable for dimensioning average-sized drawings, and for the study plates, together with the proper form of arrow heads, are given in Fig. *K*, Plate 5. The sign ′ means feet, and the sign ″ means inches: thus 10′–6½″ is read ten feet, six and one-half inches. Inclined signs are always used by printers, but in dimensioning a drawing vertical signs should be used with vertical numerals, and inclined signs with inclined numerals. The quantities feet and inches must always be separated by a dash (Fig. *K*, Plate 5). The numerator and the denominator of a fraction should each be balanced on the center line of the fraction taken as a whole (see the $\frac{9}{16}$, Fig. *K*, Plate 5).

43. Preliminaries to Drawing. A finished drawing is usually circumscribed by a ruled border line, *EEEE*, Fig. 64, the dimensions of which may be given,

Fig. 64.

or which must be determined from the size and arrangement of the drawings on the sheet. Outside of the ruled border a margin, *DDDD*, Fig. 64, should be laid off, and lines, *CCCC*, drawn for the boundary of the finished plate when trimmed. Therefore, to ascertain the size of the paper required for a drawing, add to the dimensions of the ruled border line twice the width of the margin, together with an allowance of extra paper for thumb-tack holes, which must fall outside the trimmed edge.

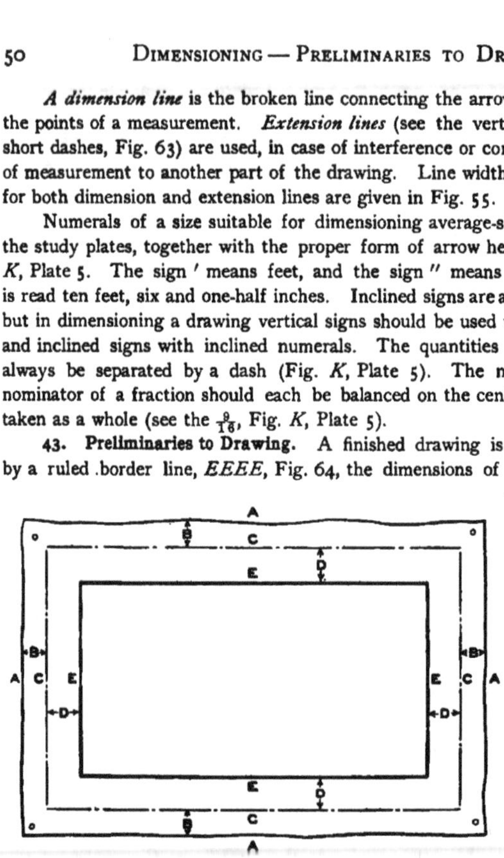

Never begin work on paper larger than the drawing board; if the paper projects beyond the board, immediately trim the edges so that all shall lie at least ⅛ inch inside the edges of the board. As the T-square is likely to work less accurately near the lower edge of the board, the paper, when smaller than the board, should be placed well above its lower edge (Fig. 66). Smooth the paper flat, place it squarely on the board, and start the thumb tacks at right angles to the board (A, Fig. 65),

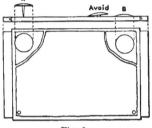

so that, when the tack is pressed in, the head will bear evenly on the paper (B, Fig. 65). *Never use the T-square to drive in thumb tacks.*

Provide a piece of clean cloth or paper with which to protect the drawing; *when not at work on the drawing, keep it covered.* Also, when working on a large or carefully executed drawing, cover all parts not receiving immediate attention.

Fig. 65.

(a) *To lay out a ruled border line.* Take, for example, a plate which shall have an 8 in. x 12 in. ruled border line and a margin 1 in. wide. Let A, A, Fig. 66, represent the edges of the paper cut for the drawing, according to the preceding paragraph. With the aid of the T-square, place the edges of the paper approximately parallel to the edges of the drawing board.

With the T-square and triangle, draw horizontal and vertical lines, B, B, $\frac{1}{2}$ in. or more from the edge of the paper, to allow for the trimming, and thumb tack holes. Perpendicular to the lines B, B, lay off 1 in., the given width of the margin. Draw lines, C, C, of indefinite lengths, and lay off the dimensions of the border line. Complete the border line with the

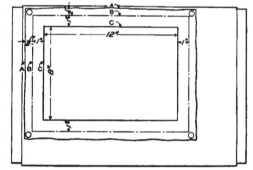

Fig. 66.

T-square and triangle. Lay off the widths of the right-hand and lower margins, and draw the remaining trimming lines.

When a strictly accurate border is necessary, it should be laid out by geometrical construction (see Problem 4, Chapter V.).

(b) *Trimming.* A very sharp knife or scissors should be used for trimming drawings. When trimmed on a board, the drawing should be so placed that the knife will be drawn across rather than with the grain of the wood; otherwise the knife is liable to follow the grain, and prevent a straight cut. The best cutting surface is thick, smooth cardboard laid on the drawing board. The regular drawing board and T-square should not be used in trimming drawings, but separate ones should be kept for this purpose. If, as a last resort, the regular drawing board and T-square are used, the cutting should be done on the back of the board and along the lower edge of the T-square.

44. Common Working Methods. (a) *Drawing by stages.* For convenience and to emphasize methodical procedure, the rendering of a drawing may be divided into stages, as, for example: (I.) *the constructive stage*, represented by the laying out of a drawing and all instrumental penciling; and (II.) *the finishing stage*, represented by the inking, or by the final lining in of a finished pencil drawing. Furthermore, a general stage may include any number of local stages, as described in connection with Figs. 69–73.

The general stages of a drawing are illustrated in Fig. 67, which represents an end of a marine engine connecting rod. The upper half of the cut shows the con-

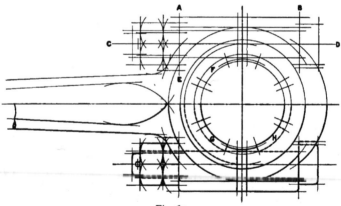

Fig. 67.

structive stage; the lower half shows the finishing stage, with the lines of the constructive stage left in for the purpose of comparison.

(b) *The constructive stage; penciling.* The penciling should invariably represent the degree of accuracy required in the finished drawing; that is, essentials must never be slighted with the idea that they may be corrected when the drawing is lined in, or inked. In the constructive stage all lines should be *full*, of uniform width, light but firm, very narrow, and made with as hard a pencil as the paper used will permit. Dash lines should not be used in this stage, as they cannot be made so rapidly as full lines ; furthermore, as suggested in Fig. 68, desired intersections, as at *A*, *B*, and *C*, are likely to be merely open spaces.

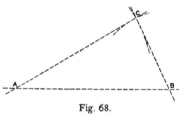

Fig. 68.

Lines upon which measurements are to be laid off, as *CD*, Fig. 67, must be drawn long enough to insure the laying off of the measurement without patching out a line. The same is the case with lines required to be intersected by subsequent lines, as *AE*, *AB*, and the other produced lines in Fig. 67. To save time and to avoid patching, circles, as at *FGH*, which are more or less broken up in the finished drawing, should be drawn complete in the penciling.

There should be little or no erasure when a drawing is in progress, and it should not be cleaned *until finished*, as the paper soils more quickly after the rubber has been used.

(*c*) *The finishing stage of a pencil drawing (finished rendering)*. The use of line conventions should be confined to this stage of the drawing, and rendered in connection with the lining in. Use a rather soft pencil (F to 3H) and emphasize all lines strongly. Dashes should be drawn with a deliberate stroke, not merely touched in, and the ends of each dash should be clearly defined. All dashes of the same convention should be equal in length, and the spaces between the dashes made equal. The lines of the constructive stage need not be erased between the dashes, as they become inconspicuous if the dashes are sufficiently emphasized. In a carefully rendered drawing, dimension and extension lines (Art. 42) should first be lightly ruled, in full line, in connection with the suggestion of the numerals, and the dashes should be put in later along with other conventions.

(*d*) *The finishing stage of an inked drawing (finished rendering)*. This stage includes, besides inking, the rendering of dimensions and lettering (Arts. 39 and 42) — which should not precede the inking — together with any penciling connected therewith. All line conventions should be rendered directly in ink ; that is, without a preliminary penciling of the convention. Make all lines perfectly smooth, and, except in the case of curved shade lines, keep all lines of the same class uniform in width. Each dash should have the same width throughout, and the ends should be made square or perpendicular to the direction of the dash.

Every line should be carried accurately to its destination, neither falling short of nor extending beyond it; special attention should be given to the rendering of corners, that all may be perfect.

(*e*) *Inking by stages.* To save time and also to minimize the chance of smearing wet ink, *similar operations should be grouped,* as indicated in the following model, which shows the steps actually taken in inking and dimensioning the original drawing for Fig. 69.

Beginning at the upper left-hand corner of the drawing and — to avoid wet ink — working downward and from left to right : (I.) Ink all circles and arcs of the

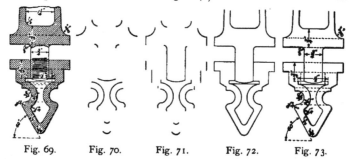

Fig. 69.   Fig. 70.   Fig. 71.   Fig. 72.   Fig. 73.

same radius, then all remaining circles and arcs (Fig. 70). (II.) Ink the vertical lines (Fig. 71). (III.) Put in all the horizontals and other remaining full lines (Fig. 72). (IV.) Draw the dash lines (Fig. 73). (V.) Render the shade lines (Fig. 73). (VI.) Render the dimensions (Fig. 73). (VII.) Draw the screw-thread convention; indicate the breaks; and cross hatch the section (Fig. 69).

(*f*) *The inking of convergent lines.* To prevent lines from running together

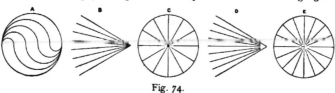

Fig. 74.

near their point of convergence, or intersection (*A*, *B*, and *C*, Fig. 74), let each line dry before inking another, and carry the pen away from the point of convergence rather than towards it; or terminate the interior lines (*D* and *E*, Fig. 74) at an arc described, in pencil, from the point of convergence taken as center.

(*g*) *Shade lines.* When shade lines (Art. 58) are shown on a drawing, the extra width of line, whenever practicable, should be added to the outer edges of straight and curved lines so as not to encroach upon the surface area bounded by the lines. In the case of a circle or a circular arc, the shade line should be placed by shifting the center. Thus, for example, in drawing shade lines *mhk* and *egf*, Fig. 75, the center was shifted from *a* to *c*, a point in the line *dg* drawn at 45° to the horizontal, the distance *ac* being determined by eye. It will be observed that the shade lines, as thus drawn, do not encroach on the surface included between the two circles. Likewise the shade line *stuv*, added to the outer edge of line *st*, does not encroach upon the surface, *qrst*, of the ring.

Fig. 75.

(*h*) *Testing.* The accuracy of a drawing should be frequently verified by checking or testing ; in office practice, this is the only safeguard against costly mistakes.

45. Solution of Geometrical Problems by Practical Working Methods. It has already been stated (Art. 23) that in practical drawing parallels, perpendiculars, and angles of 15°, 30°, 45°, 60°, and 75° are obtained by means of the T-square and triangles, or by the triangles alone. There are many other cases, however, where exact geometrical construction (see Chapter V.), is unnecessarily laborious, and where accurate results may be obtained by shorter methods, some of which are given below. Apart from the value of the methods as such, it is intended that they shall suggest to the student further possibilities in the use of the triangles and compass. Speed, without the sacrifice of accuracy, often depends on a ready application of some particular instrumental method.

(*a*) *To draw a line perpendicular to a given line at its middle point.* Let *AB* be the given line. From the ends, *A* and *B*, of the line, draw lines *AC* and *BC*, making equal angles with *AB*. (Make these equal angles either 30°, 45°, or 60°; if the given line is neither horizontal nor vertical, proceed according to *D* and *E*, Fig. 23.) From the intersection *C*, draw the required line *CD* perpendicular to *AB* (see *B*, Fig. 23).

(*b*) *To draw a circular arc through three given points.* Let *A*, *B*, and *C* be the given points. Draw *AB* and *BC*. Draw *DO* perpendicular to *AB* at its middle point (see *a*). Draw *EO* perpendicular to *BC* at its middle point. The intersection, *O*, is the center of the required arc.

(*c*) *To draw a tangent to a circular arc at a given point on the arc.* Let *C* be the given point on the arc, center *O*. Draw the radius *OC;* at *C* draw the required tangent *AB* perpendicular to *OC* (*B*, Fig. 23).

(*d*) *To find the point of tangency of a given straight line and circular arc.* Let *AB* be the tangent to the arc, center *O*. From *O* draw *OC* perpendicular to *AB* (*B*, Fig. 23); the intersection, *C*, is the required point of tangency.

(*e*) *To find the point of tangency of two given circular arcs.* Let *A* and *B* be the centers of the given arcs. Draw *AB;* the intersection, *C*, is the required point of tangency.

(*f*) *To draw an arc of a given radius tangent to two given lines at right angles.* Let *AB* and *BC* be the given lines. Set the compasses to the given radius; then, with *B* as center, intersect *AB* and *BC* at *D* and *E*. With the same radius, centers *D* and *E*, draw arcs intersecting at *O*, the center of the required arc. Points *D* and *E* are the points of tangency.

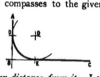

(*g*) *To draw a line parallel to a given line at a given distance from it.* Let *AB* be the given line. Set the compasses to a radius equal to the given distance, then, with any point on *AB*, as *C*, for a center, draw an arc. Draw, tangent (by eye) to the arc, the required line *DE* parallel to *AB* (*A*, Fig. 23). *Note.* There are two solutions possible, one on each side of the given line.

(*h*) *To draw a circular arc parallel to a given circular arc, and at a given distance from it.* Let *AB*, center *O*, be the given arc. Draw any radius *OC*. Make *CD* equal to the given distance. With radius *DO*, center *O*, draw the required arc *DE*. *Note.* There are two solutions possible, one on each side of the given arc.

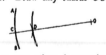

(*i*) *To draw an arc of a given radius tangent to two given intersecting straight lines.* Let *AB* and *BC* be the given lines. Draw *DE* parallel to *AB* at a distance equal to the given radius (see *g*). Draw *FG* parallel to *BC* at a distance equal to the given radius. The intersection, *O*, is the center of the required arc; the points of tangency may be found according to *d*. *Note.* There are four solutions possible.

(*j*) *To draw an arc of a given radius tangent to a*
given circular arc and to a given straight line. Let *AB*
be the given circular arc, and *AC* the given straight
line. Draw arc *DE* parallel to arc *AB* at a distance
equal to the given radius (see *h*). Draw line *FG* parallel
to line *AC* at a distance equal to the given radius (see *g*). The intersection, *O*,
is the centre of the required arc; the points of tangency may be found according
to *d* and *e*. *Note.* There may be four solutions possible.

(*k*) *To draw an arc of a given radius tangent to*
two given circular arcs. Let *AB* and *BC* be the given
circular arcs. Draw arc *EF* parallel to arc *AB* at a distance equal to the given radius (see *h*). Draw arc *GH*
parallel to arc *BC* at a distance equal to the given
radius. The intersection, *O*, is the center of the required arc; the points of
tangency may be found according to *e*. *Note.* There may be four solutions
possible.

(*l*) *To bisect a given angle.* Let *ACB* be the
given angle. Make *CD* and *CE* any equal distances.
Draw *DE*. From *C* draw *CF*, the bisector, perpendicular
to *DE* (*B*, Fig. 23).

(*m*) *To draw an arc tangent to three given*
straight lines. Let *AB*, *BC*, and *CD* be the given
lines. Bisect the angles *ABC* and *BCD* (see *l*). The
intersection, *O*, of the bisectors is the center of the
required arc.

(*n*) *To draw an arc tangent to two given straight lines, and to a circle the*
center of which lies on the bisector of their angle. Let *AB* and *BC* be the given
lines, *DB* the bisector of the angle *ABC*, and *F*, lying
on *DB*, the center of the given circle. At *E*, in the
given circle, draw the tangent *GH* perpendicular to *DB*
(see *c*). Draw the required arc, center *O*, tangent to
lines *AG*, *GH*, and *HC* (see *m*).

(*o*) *To draw an arc tangent to two equal circles, and passing through a*
point equally distant from their centers. Let *A* and *B* be the centers of the given
circles, and *P* the given point. Draw, if not already
given, line *PC* perpendicular to *AB* at its middle point.
Make *PC* equal to the radius of either circle. Find
the center, *O*, of an arc which would pass through points
A, *C*, and *B* (see *b*). Point *O* is the center of the required arc *EPF*.

(*p*) *To draw a series of arcs of constant radius, tangent to a series of equal circles.* Find the center *O* according to methods *k* or *o*. Since the given circles are equal, *AO* equals *BO*. Using the constant radius *AO*, find all the centers *O'*, *O''*, etc.; then, using the constant radius *CO*, draw all the required tangent arcs.

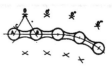

(*q*) *To draw a circular arc tangent to two given straight lines, and passing through a given point.* Let *AB* and *BC* be the given lines, and *P* the given point. Draw the bisector, *BD*, of the angle *ABC* (see *l*). Find the required center, *O*, on the line *BD*, by trial. *Note.* This is typical of cases where one line through the required center may readily be drawn, but the remainder of the geometrical construction is too complicated to be easily remembered.

(*r*) *To draw a circular arc tangent to two given circles, and passing through a given point.* Let the circles be described from the centers *A* and *B*, as shown, and let *P* be the given point. Find the center, *O*, of the required arc by trial. *Note.* This is typical of cases where the entire geometrical construction is too complicated to be easily remembered.

(*s*) *To draw a regular hexagon, given its short diameter.* Let *AB* be the given diameter. On *AB* as a diameter draw a circle. Using the T-square and 30°–60° triangle, draw the sides of the required hexagon tangent (by eye) to this circle. *Note.* A regular octagon may be similarly constructed, using the T-square and 45° triangle.

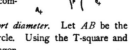

(*t*) *In a scale drawing, to find the radius of a circle, given its diameter.* Lay off the length of the given diameter from the scale which is one-half the given scale; this will be the required radius. *Note.* This is a convenient method for dividing any scale measurement by 2 without figuring; thus, if a dimension be laid out from the scale of 3″ = 1 ft., then the same dimension taken from the scale of 1½″ = 1 ft. will be one-half as long.

CHAPTER IV.

STUDY PLATES ON INSTRUMENTAL RENDERING AND CONSTRUCTION.

46. As the ability to exercise judgment in securing speed depends upon first grasping the leading points of the work as a whole, in the following study plates the student should not proceed piecemeal — merely drawing as he reads — but should read all the directions before beginning to draw (see the last paragraph, Art. 35).

STUDY PLATE 1.

For accuracy and speed in the use of the T-square, triangles, scale, pricker, and ruling pen; testing; tracing; rendering of letters and dimensions.

Use duplex detail paper. The size of the finished plate is to be 14″ x 20″. It is required to make a drawing for a tracing on cloth.

I. PENCILING. Sharpen the 4H pencil to a ruling point (*a*, Art. 19). Make all the lines *full*, very narrow and light, but sufficiently distinct to be readily seen through the tracing cloth. Lay out a 12″ x 18″ border line; work from the upper and left-hand edges of the paper (*a*, Art. 43). For location and other measurements, see Plate 6.

(*a*) Locate, and rule with the T-square, the horizontal lines, Fig. *A*, Plate 6; start each of these lines at a vertical line, *PQ*, (shown also at *KL*, Fig. 76), drawn 1⅛″ from the left-hand side of the border line. Using the scale and the pricker, lay off accurately (*e*, Art. 36), as many times as the length of each line will permit, the following measurements: on line *A*, ⅞″; on *B*, $\frac{7}{16}$″; *C*, $\frac{7}{32}$″; *D*, ¾″; *E*, ⅞″; *F*, $\frac{3}{16}$″; *G*, $\frac{3}{32}$″. On line *H* lay off in succession $\frac{9}{32}$″, $\frac{15}{32}$″, $\frac{3}{16}$″, $\frac{9}{16}$″, ⅜″, $2\frac{1}{2}$″, and $\frac{9}{16}$″. In laying off the measurements on each line, *do not move the scale.* Test. Using T-square and triangle, drop a perpendicular, as *ac*, Fig. 76, from each point of measurement in line *A*. See whether each perpendicular passes through the imaginary center of

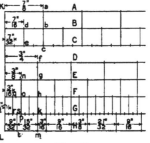

Fig. 76.

each alternate division in line *B*, as point *b*; also, whether each perpendicular from line *A* passes through every fourth point, as *c*, in line *C*. Continue the test as shown by Fig. 76.

(*b*) With the triangle placed against the T-square, rule the vertical lines, Fig. *B*, Plate 6.

(*c*) Draw the rectangles, Fig. *C*, to the scales indicated.

(*d*) Locate point *A* in Figs. *D*, *E*, *H*, and *J*, and point *B* in Figs. *F* and *G*. Through these points draw, indefinite in length, the vertical and horizontal lines. Rule the equally spaced parallel lines, making the angles of 45°, 60°, and 30° with the horizontal; in each case, the spaces must be laid off on a line, *AB*, drawn at right angles to the required lines (see Fig. 20).

(*e*) Draw the line *AB*, Fig. *K*, and locate point *C*. Draw the equally spaced lines parallel to *AB* (see *A*, Fig. 23). Rule the lines perpendicular to *AB* (see *B*, Fig. 23), and lay off their length. Draw the lines from points *E* and *F* (see Fig. 23); terminate the lines by circular arcs, as shown in Fig. *K*.

(*f*) Using the T-square and 30°–60° triangle, draw Figs. *L* and *M*. *Test.* In Fig. *L*, with the 30°–60° triangle placed against the T-square, bisect the angles, and see if the bisectors intersect in the same point. In Fig. *M*, with the 45° triangle placed against the T-square, draw the diagonals of the square, upward from the ends of the base, and see if each passes accurately through an upper corner of the square.[*]

(*g*) Begin Figs. *N* and *O* by repeating Figs. *L* and *M*. Find, by scale measurement, the middle point of each side of the triangle, Fig. *N*, and draw *CD*, *BD*, and *AD*, respectively, perpendicular to a side of the triangle. Lay off on each perpendicular the measurements given on *BD*. Through these points, using T-square and triangle, draw the sides of the interior triangles. *Test.* Produce *CD*, *BD*, and *AD*, and see if the alignments of the corners of the triangles are accurate.

Find by scale measurement the centers *C* and *B* of two sides of the square, Fig. *O*. Draw the diameters, *CD* and *AB*, of the square. On each semi-diameter lay off the measurements given at *C*. Through these points draw the sides of the interior squares. *Test.* Draw the diagonals of the outer square and see if they pass through the center of the square, as located by the diameters, and also if they pass through the corners of the inner squares.

(*h*) Draw the hexagon, Fig. *P*, according to the given angles and measurement. *Test.* Connect the opposite angles of the hexagon, and see if the diagonals thus obtained intersect in the same point.

(*i*) Draw *AB*, Fig. *Q*. Using T-square and triangle, complete the hexagon without further scale measurement. *Test.* Measure with the scale each side of the hexagon, and see if all have the same length.

(*j*) The student's name, the date, and all lettering and dimensions on Plate 6 are to be given on the tracing, and rendered with strokes according to Art. 40. As

---

[*] If the results are not accurate, it should be found whether the fault lies in the triangle (see *a*, *b*, and *c*, Art. 6).

Plate 6
(Study Plate I)

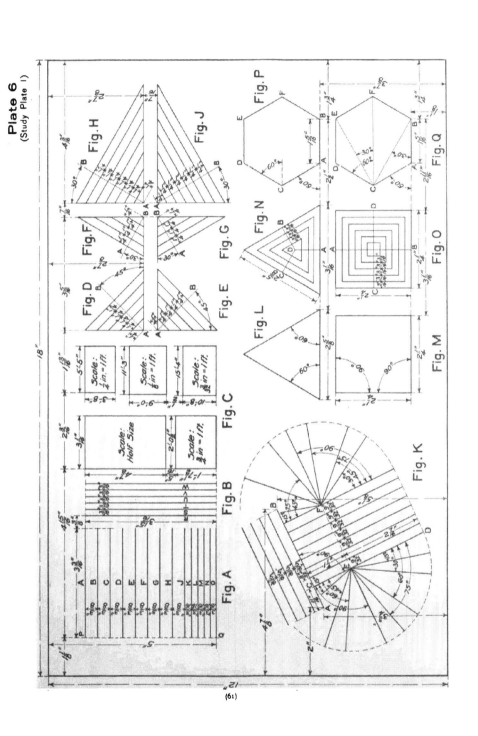

a very general guide for the stroke letters and numerals on the tracing, suggest, lightly and rapidly, their position, size, and spacing ; one guide line may be ruled for each line of letters and for each dimension. Use Figs. *A* and *F*, Plate 5, as a guide, but determine the sizes of the letters and numerals wholly by eye.

II. TRACING. Work on the dull surface of the tracing cloth. Smooth out the cloth as flat as possible, and fasten with four additional thumb tacks placed midway between the corners. See that the ink is *black*. If the cloth does not take the ink well, use chalk (Art. 17). The lines made on the tracing cloth should cover the lines in pencil accurately.

(*k*) Indicate the scale measurements laid off in the pencil drawing on lines *A—H*, Fig. *A*, by ruling through the center of each point of division a very narrow line perpendicular to and extending about $\frac{1}{16}''$ above and below the given line. Make the lines *A—H*, Fig. *A*, full, and of the width given in *A*, Fig. 55. Make the lines *J—O*, Fig. *A*, to correspond to the lines *D—J*, Fig. 55 ; repeat for the lines *R—W*, Fig. *B*. In the remaining figures make all result lines the same as the line *D*, Fig. 55 ; make incidental lines, as *AB*, Figs. *D—J*, Plate 6, like *A* or *B*, Fig. 55. The line of the ruled border should be slightly wider than the result lines in the drawing. *Test.* Measure with the scale the distances on lines *A—H*, Fig. *A*, as established in the tracing ; see if they correspond with the measurements given in Fig. 76.

(*l*) Letter and dimension the tracing in the following order : —

Render, directly in ink, with strokes of the pen (*a*, Art. 40), the lettering, dimensions, and signs for inches ; rule the dimension lines ; put in the extension lines ; render the arrow heads. Letter with care, but rapidly.

(*m*) Erase the lines radiating from *E* and *F*, Fig. *K*, (*d*, Art. 33) ; also, the dimensions locating the points *E* and *F*, and the letters *E* and *F*. Use erasing shields to protect the rest of the figure as much as possible. Relocate points *E* and *F* at $1\frac{5}{8}''$ from line *AB*. Move the tracing cloth to correspond with this new location, and retrace the lines radiating from *E* and *F*, terminating these lines on arcs drawn with a radius equal to the new length of *ED*. Reletter *E* and *F ;* correct and put in the dimensions for the new location of these points. Restore any lines injured in making these alterations.

(*n*) Hand in both the pencil drawing and the tracing. *Do not roll or fold the drawings.*

STUDY PLATE 2.

For accuracy and speed in the use of the compass, dividers, and French curve ; testing ; tracing ; the rendering of letters and dimensions.

Read all of the following directions before beginning to draw.

Use duplex detail paper. The size of the finished plate is to be 14″ x 20″, the ruled border line 12″ x 18″. It is required to make a drawing for a tracing on cloth. For location and other measurements see Plate 7.

I. PENCILING. Use the 4H pencil. Make all lines *full*, very narrow, and light, but sufficiently distinct to be readily seen through the tracing cloth.

(*a*) Describe the circle *A*, Fig. *A*. Rule a line from the center to the circumference, and on this line lay off the spacing of the interior circles, the smallest of which is $\frac{1}{8}$″ diameter; describe the circles according to *a*, Art. 27. From the same center, using the lengthening bar (*b*, Art. 27), draw the arcs, Fig. *C*.

(*b*) Work the following steps very accurately. Draw *AB*, Fig. *B*, and locate the center, *C*, by scale measurement. Using only the T-square and the 30°–60° triangle placed against the T-square, draw the hexagon, its diameters, and its diagonals. Distant $\frac{5}{8}$″ from *A*, locate point *H*, and, with *C* as center, describe the circle *OHK*. With $\frac{5}{8}$″ radius, centers on circle *OHK*, describe the circles tangent to the sides of the hexagon. Establish the points of tangency, as *P*, *Q*, and *R*, by drawing, with *C* as center, the circle through point *P*, which is the intersection of *HK* and *CT*. *Tests*. With *C* as center, radius *CA*, describe a circle; see if the imaginary center lines of the three tangent lines at *A*, *G*, *E*, *B*, *F*, and *D*, intersect in a point (Fig. 52). With $\frac{5}{8}$″ radius, *C* as center, draw a circle; see if, in its intersection with each diameter of the hexagon, it is tangent to each of the six equal circles. If the results in the foregoing constructions are found to be inaccurate, all lines should be erased, the line *AB* moved $\frac{1}{8}$″ to the left, and the construction repeated.

(*c*) From line *LH*, indefinite in length, draw lines *A*—*F*, Fig. *D*. With the hair-spring dividers, and according to *a*, Art. 29, and *f*, Art. 36, space the lines into equal parts, as follows: *D*, 17 parts; *B*, 13; *C*, 11. With the bow spacers divide equally as follows: *A* into 17 parts; *E*, 13; *F*, 11. *Test*. Locate point *L* by producing line *JK*. Draw through point *L* a straight line from each point of division on line *D*. See if the line from each of these points passes through the corresponding point of division in line *A*. Test in like manner the points of division in lines *B* and *E*, and in lines *C* and *F*.

(*d*) Draw, Fig. *E*, the circles, circular arcs, and line *EM*. With the hair-spring dividers, and starting at line *EM*, space the circles *A* and *B* each into 19 equal parts. With the bow spacers, starting at line *FM*, divide circles *C* and *D* each into 19 equal parts. *Test*. Draw very accurately from center *E* to each point of division in circle *A*; see if each line passes through the centers of the corresponding divisions in circles *B*, *C*, and *D*.

Draw *EG* through point *14*, Fig. *E*. With the hair-spring dividers space equally arc *HN* into 5 parts; *JO* and *LQ* each into 9; *KP* and *MG* each into 7 parts. *Tests*. Draw from center *E*, through points *15*, *16*, *17*, and *18*, circle *A*; see if these lines produced pass through the centers of the points of division in arc *HN*. From center *E* draw through the points of division in *JO*, and produce the

Plate 7
(Study Plate 2)

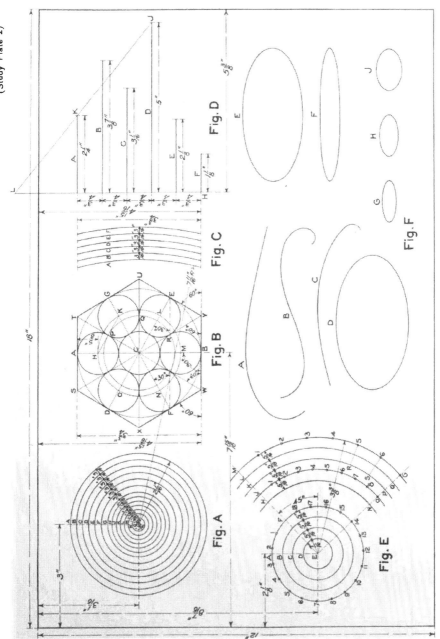

Fig. A

Fig. B

Fig. C

Fig. D

Fig. E

Fig. F

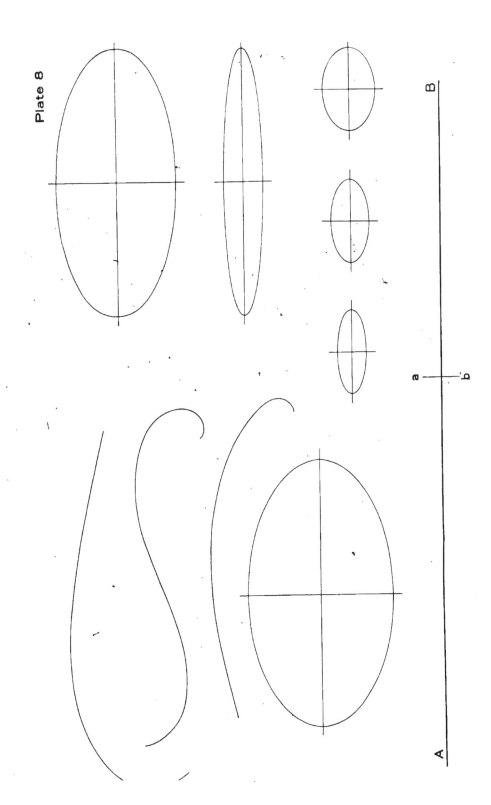

Plate 8

lines to intersect LQ; see if the lines pass through the centers of the points of division in LQ. From center E draw through the points of division in MG; see if the lines pass through the centers of the points of division in KP.

(*e*) The curves A—J, Fig. F, should be located by pricking through the curves given in Plate 8, and the points thus obtained are to be connected freehand (*a*, Art. 25). To prick off the points, proceed as follows: Draw a short perpendicular across the lower horizontal line of the ruled border line, 6″ from its right-hand end. Cut holes through Plate 8, one hole at each end of line AB, and one on ab. Now tack this plate over the drawing, taking care not to pass the thumb tacks through the drawing or its margin; place AB over the ruled border line, and ab on the perpendicular located 6″ from the corner of the border line. The location and the number of the points to be pricked through must be determined by judgment, aided by the following suggestions. In the sharp curves at the ends of curve F, Plate 7, take the points from $\frac{1}{32}$″ to $\frac{1}{18}$″ apart, and, for the flattest portions of the same curve, from $\frac{1}{4}$″ to $\frac{3}{8}$″ apart. The points in the more sharply curving portions of A, B, and C, Plate 7, should be taken from $\frac{1}{8}$″ to $\frac{3}{16}$″ apart. In the ellipses, the ends of both axes should be pricked through and the axes ruled before sketching in the curve.

(*f*) Suggest rapidly the lettering and dimensions given on Plate 7, your name, and the date, according to *j*, Study Plate 1.

II. TRACING. Work on the dull surface of the tracing cloth. See that the ink is *black*. Trace the drawing accurately.

(*g*) Ink in the circles A—F, Fig. A, to correspond to the lines A—F, Fig. 55. Repeat for the circles G—M, Fig. A. Ink the circles N and O according to line A, Fig. 55. Ink in the arcs A—F, Fig. C, to correspond to the lines A—F, Fig. 55. In the remaining figures ink in the result lines according to line D, Fig. 55, the incidental lines according to line A, Fig. 55. Omit the lines used in the tests. Indicate the points of division obtained in the spacing, by a short line perpendicular to the divided line, whether circular or straight, as directed in *k*, Study Plate 1. The line of the ruled border should be somewhat wider than the result lines in the drawing.

(*h*) Render the lettering, dimensions, dimension and extension lines, signs, and arrow heads according to *l*, Study Plate 1.

(*i*) Erase and redraw (*d*, Art. 33) the ellipse D, Fig F, but place the ellipse $\frac{1}{8}$″ to the left of the position in which it was first drawn.

(*j*) Hand in both the pencil drawing and the tracing. *Do not roll or fold the drawings.*

STUDY PLATE 3.

For the rendering of freehand lines; spacing by eye; line shading.

Read all of the following directions before beginning to draw.

Use duplex detail paper. The size of the finished sheet is to be 14″ x 20″, the ruled border line 13″ x 19″ ($\frac{1}{2}$″ margin).

Draw in pencil the outlines of the figures, Plate 4, according to the measure-ments on the plate.   Trace these outlines on cloth, and then, without further penciling, render the several conventions and line shadings, Plate 4, directly on the tracing cloth.

I. PENCILING. Use the 4H pencil.   Make all lines *full*, very narrow, and light, but sufficiently distinct to be readily seen through the tracing cloth.

(*a*) The heights of the lower line of cylinders in Fig. *B* are determined by points *a*, *b*, . . . *g*, lying in line *ag* located by measurement.   The spaces between' the lines of the cross hatching, and the width of the wider lines in examples *L*, *M*, *R*, and *S*, should be twice as great as those on the plate.   In the upper left-hand corner of each rectangle, draw not more than four or five of the lines of the cross hatching, to be used as a gage for the cross hatching on the tracing; determine the widths of these spaces and lines by measurement.

II. INKING. Work on the dull side of the cloth.

(*b*) *The conventions*.   After tracing the outlines of the figures, read Art. 38 and practice each convention on spare tracing cloth before rendering it on the final drawing.   As the cross hatching proceeds, compare frequently the widths of the spaces and lines with those of the gage.

(*c*) *Lettering*.   Letter "Plate 3," your name, and the date.   Rule guide lines, and draw (not stroke render) the letters (Art. 41).

(*d*) Hand in both the pencil drawing and the tracing.   *Do not roll or fold the drawings.*

STUDY PLATE 4.

For accurate spacing, scale measurement, and penciling ; the drawing of tangent circular arcs ; finished rendering in ink ; lettering and dimensioning.

Read all of the following directions before beginning to draw.

Use Whatman's hot-pressed paper.   The size of the finished plate is to be 14" x 20", the ruled border line 12" x 18".

It is required to make a full-size drawing of the bicycle chain and sprockets, Plate 9.   To obtain a satisfactory arrangement, it is necessary to break the chain and to place the sprockets $10\frac{1}{4}$" apart instead of 18", the actual distance.   The dia gram, Fig. *E*, gives the center line of the complete chain, and Fig. *F* supplements the description for drawing the sprocket teeth and links of the chain.

I. PENCILING. Use the 6H pencil.   Make all lines *full* and very narrow.

(*a*) Lay out the ruled border line.   Draw the horizontal center line, *OQ*, and the vertical center lines, *VW* and *XY*, located according to the measurements on the plate.   With radius $3\frac{11}{32}$",* center *P*, draw the pitch circle, *J*, of the front

---

* The diameters $6\frac{11}{16}$", Fig. *A*, and $2\frac{13}{16}$", Fig. *B*, are the nearest fractional equivalents of the true (decimal) pitch diameters given on the plate (see table at the end of the book).

Plate 9
(Study Plate 4)

BICYCLE CHAIN AND SPROCKETS

SCALE FULL SIZE

Fig. A

Fig. B

Fig. C

Fig. D

Fig. E

Fig. F

Fig's E and F and dimensions marked * should not appear on the student's drawing

(71)

sprocket, Fig. *A*. Draw line *GH* tangent to circle *J*. Tangent to *GH*, draw the pitch circle, *N*, of the rear sprocket, Fig. *B*. (Lay off the diameter, $2\frac{6}{14}''$, on the centre line *XY*, and find the centre by bisecting).

(*b*) *To find the direction of the lower portions, TU and RS, of the chain.* Tack a piece of paper over the drawing so as not to cover the lower half of the circles *J* and *N*. (Do not place the tacks inside the boundary of the finished plate.) On this extra sheet make a half-size diagram of the center lines, similar to Fig. *E*, placing line *G'H'* parallel to *GH*. Draw the common tangent, *T'S'* ; then, parallel to *T'S'*, draw the required tangents, *TU* and *RS*. (Do not draw this diagram directly on the finished drawing, as the necessary erasure will mar the paper.)

(*c*) *The teeth of the sprockets and the links of the chain.* It will be seen (Figs. *C* and *F*) that the chain is composed of alternate closed and open links, and that the teeth mesh into the latter. The inside length of an open link determines the width, *Y'Z*, Fig. *F*, of the teeth, while the shape of the interior determines the curves *X'*, *X'*, of the teeth. The spaces, *ZY'*, between the teeth, must equal the total length of a closed link.

Construction: the teeth of the front sprocket. Starting at point *G*, Fig. *A*, with the hair-spring dividers space the pitch circle accurately into 21 equal parts (*a*, Art. 36), and, to identify the points, enclose each point in a small freehand circle. Let the first two spaces, *G''V'* and *V'V''*, Fig. *F*, be typical of all of the 21 spaces. Take in the bow spacers the given distance between the centers of an open link (that is, $\frac{9}{16}''$), and lay it off from points *G''* and *V'*; this establishes the centers of the pins of all the links. With $\frac{5}{32}''$ radius, centers *G''*, *W'*, *V'*, *W''*, and *V''*, draw (all the way around the pitch circle) complete circles representing the ends of the links; these circles determine the curves *X'*, *X'*, of the teeth. With radius $\frac{13}{32}''$ ($\frac{9}{16}''$ minus one-half of $\frac{5}{16}''$) describe, from the centers of the pins, the curves *Y'* and *Z* of the teeth. The extremities of the teeth and the bottoms of the spaces are portions of the circles *abc* and *def*, Fig. *F*, which are drawn according to the measurements given in Fig. *A*.

The teeth of the rear sprocket. Starting at point *H*, Fig. *B*, divide the pitch circle accurately into 8 equal parts. Draw the teeth according to the directions given for the teeth of the front sprocket.

The chain. The pins and ends of the links for the portions of the chain in contact with the sprockets are already drawn. Locate the centers of the pins for the straight portions of the chain, and draw complete circles determining the ends of the links. Find the radii for the sides of the links (*o*, Art. 45); find all the centers for arcs that have the same radius, and draw all such arcs as one operation (*p*, Art. 45).

(*d*) Draw the circles containing the centers of the holes in the webs of the sprockets, and space each circle for the number of holes shown.

(*e*) Locate Fig. *C* and Fig. *D*, according to the measurements given on the plate. Locate all lines in Figs. *C* and *D* not determined by the given dimensions, by projecting from Figs. *A* and *B*. The chain, Figs. *C* and *D*, is projected from the horizontal portions of the chain, Figs. *A* and *B*.

(*f*) Complete the penciling. Look over the drawing, and see that no details have been omitted. Do not indicate the dimensions and lettering before inking the drawing.

II. INKING. (*g*) Ink the drawing with special care; make all lines at first of the same width, and equal to one-half that of the line *D*, Fig. 55.

(*h*) *Shade lines.* Add the shade lines; make their width equal to three-quarters that of the line *E*, Fig. 55, and render them according to *g*, Art. 44. (Also see Art. 58.)

(*i*) *Lettering and dimensions.* All dimensions given on Plate 9, except those marked *, should appear on the student's drawing. The letters and numerals should be drawn (Art. 41), not stroke rendered. Rule guide lines according to Figs. 58 and 62; make the size of the letters and numerals the same as in Fig. *E*, Plate 5. Rule guide lines for the title, according to the measurements given on the plate, and balance the title on the vertical center line of the plate (*b*, Art. 41). Letter " Plate 4," your name, and the date.

(*j*) Erase pencil lines throughout the drawing.
Do not roll or fold the drawing.

STUDY PLATE 5.

For precise spacing with the dividers and bow spacers; strictly accurate scale measurement and penciling; finished rendering in ink; lettering, and dimensioning.

Read all of the following directions before beginning to draw.

Use Whatman's hot-pressed paper. The size of the finished plate is to be 10" x 14", the ruled border line 8" x 12".

It is required to draw the spur gear, Plate 10. The " *Scale half size,*" on the plate, refers to the scale of the drawing. As the measurements on a drawing indicate the actual size of the object, it is evident that in making this drawing all measurements on the plate must be divided by 2, except the dimensions marked *. The diagram, Fig. *C*, is given to supplement the directions which follow, and should not appear in the student's drawing; note that corresponding lines in Figs. *A* and *C* are similarly lettered.

I. PENCILING. Use the 6H pencil. Make all lines *full* and very narrow.

(*a*) Lay out the border line and draw the center lines *MO* and *DP*, located as shown on the plate. With point *N* as center, radius $3\frac{8}{8}''$ (equals $13\frac{1}{2}''$ divided by 4), draw the *pitch circle G* (*G'*, Fig. *C*).

(*b*) *Dividing the pitch circle.* Starting at point *a*, Fig. *C*, lying in the

Plate 10
(Study Plate 5)

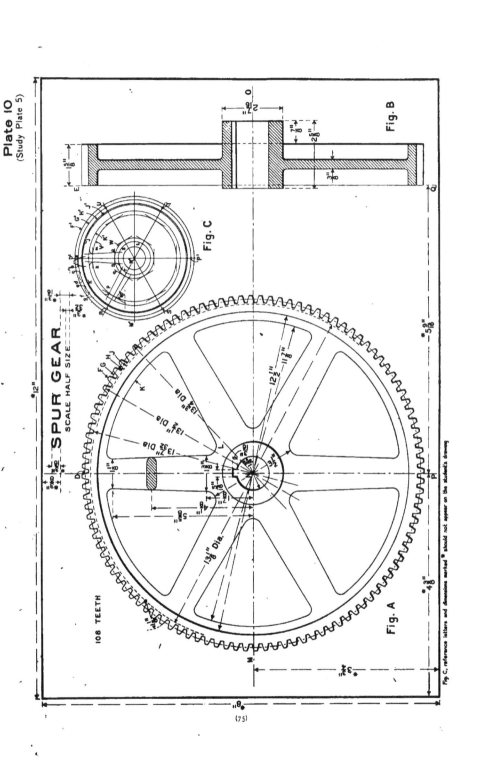

SPUR GEAR
SCALE HALF SIZE

108 TEETH

Fig. A

Fig. B

Fig. C

Fig. C, reference letters and dimensions marked * should not appear on the student's drawing

center line $D'P'$, divide with great precision (f, Art. 36) the pitch circle into 108 equal parts, each to contain one tooth and one space (between adjacent teeth). As it is next to impossible to divide a circle directly into so great a number of equal parts, first divide the circle very accurately into 6 equal parts (see points a, R, S, P', T, and U, Fig. C); for this, do not use the 30°–60° triangle, but lay off the distances with the hair-spring dividers set to the radius of the pitch circle. To see that the measurement is exact, test the division by stepping the dividers in an opposite direction before indenting the points; to identify the points, draw very lightly a small freehand circle about each. Using the bow spacers, divide each of the six divisions into 18 equal parts; test the divisions by restepping in an opposite direction; draw a freehand circle about each point of division.

(c) *The width of the teeth.* Let ac and ce, Fig. C, represent the first two of the 108 spaces. In the present drawing the widths of the teeth are made equal to the spaces between them. Take one-half of ac in the bow spacers, and, starting at point a, lay off this distance from each of the points marking the 108 spaces,—as shown for the teeth at ab, cd, and ef. In order that the space divisions may not be mistaken for teeth divisions — not an uncommon mistake — each tooth division should be identified by a very light freehand line, as shown between c and d, Fig. C.

(d) *The outline of the teeth.* The extremities of the teeth and the bottom of the spaces are circular arcs; in the present case the sides of the teeth are also circular arcs. The centers of the sides of the teeth lie in the circle J. Draw circles F, H, and J, according to the measurements in Fig. A. To complete the teeth in pencil: With the same radius throughout (namely, one-half of $1\frac{1}{8}''$), center at the point marking a side of a tooth, as a, Fig. C, cut the circle of centers, J', at point h. With center h, draw the side of the tooth through point a. With center at the point b, marking the opposite side of the tooth, cut the circle of centers at point j. With center j, draw the side of the tooth. Draw each tooth in a similar manner.

(e) *The arms of the gear.* Draw circle K', and the center lines RT and SU of the arms. The widths of each arm are measured in lines, perpendicular to the center line of the arm, located from the center of the gear. With radii equal to one-half of $5\frac{5}{8}''$, and one-half of $1\frac{1}{8}''$, respectively, draw the circles V and W. Tangent to these circles, draw the lines upon which to lay off the widths of the arms. (Use the 30°–60° triangle to draw the perpendiculars to the center lines RT and SU.) Lay off the arm widths, and draw the sides of the arm produced, as kn and mo.

Make the circular arcs (fillets) connecting adjacent sides of the inner ends of the arms, of $\frac{3}{16}''$ radius (not given on the plate); draw them according to n, Art. 45. Make the fillets, at the outer ends of the arms, of $\frac{1}{8}''$ radius (not given on the plate); draw them according to j, Art. 45.

(*f*) *The section of the gear*, Fig. *B*, is taken on line *DP* (Fig. *A*). Locate line *EQ* as shown on the plate. All parts not determined by the given dimensions must be projected from Fig. *A*. Make the radius of the fillets $\frac{1}{8}''$ (not given on the plate); draw them according to *f*, Art. 45.

(*g*) Complete Fig. *A* : the diameter of the hub must be taken from Fig. *B*. Draw the cross section of the arm as shown, Fig. *A*, taking its width from the section, Fig. *B*.

(*h*) Do not cross hatch the section, or indicate the measurements and lettering before inking the drawing.

II. INKING. (*i*) Ink the drawing with special care; make all lines at first of the same width, and equal to one-half that of the line *D*, Fig. 55.

(*j*) *Shade lines*. Add the shade lines; make their width equal to three-quarters that of the line *E*, Fig. 54; render them according to *g*, Art. 44. (Also see Art. 58.)

(*k*) *Lettering and dimensions*. All dimensions given on Plate 10, except those marked *, should appear on the student's drawing. The letters and numerals should be drawn (Art. 41), not stroke rendered. Rule guide lines according to Figs. 58 and 62; make the size of the numerals the same as in Fig. *E*, Plate 5. Rule guide lines for the title, according to the measurements given on the plate, and balance the title on the vertical center line of the plate (*b*, Art. 41). Letter "Plate 5," your name, and the date.

(*l*) Erase pencil lines throughout the drawing.

Do not roll or fold the drawing.

CHAPTER V.

GEOMETRICAL CONSTRUCTION.

47. The following problems are given for further practice in precise rendering and for their practical application. It should be understood that the use of such problems is not necessarily confined to the drawing room or to drawings of the usual sizes. A good example of this may be found in surveying, where problems are often worked out on a large scale, the straight lines being run out with the transit, which corresponds to the straight-edge in drawing, and the circular arcs swung with the chain, or tape as a substitute for the drawing compass. Likewise, the landscape gardener lays out geometrical figures directly on the ground, stretching the tape for a straight-edge, and describing arcs with the aid of tape and measuring pins. In the mechanic arts, the workman may need to lay out a construction to the actual size of his work, on wood, metal, or the floor of the shop; in large work substituting chalk line for straight-edge, and striking the arcs with a piece of chalk held at the end of a string or a strip of wood swung from a nail as center.

PENCILING. Use a 6H pencil, make all lines *full*, and draw with the greatest accuracy possible. For appearance, intersecting arcs should be made of equal length, and should be at right angles.

Instead of always following a geometrical method to its limit, it is better in some cases to rely upon instrumental methods. For example, parallels (as in Prob. 13) may be drawn by sliding the triangles (Art. 23); circular arcs may be divided (as in the case of D5B, Prob. 27) with the bow spacers; and horizontals and perpendiculars (as in Prob. 44) may be drawn with the T-square and triangle.

When practicable, final results should always be checked: for example, in Prob. 13, see that $AF = BE$; in Prob. 24, see whether $AD = CE = CB$.

INKING. Either of the following systems may be used: —

(*a*) All lines to be in *black*. Given lines, *dash and dot*; construction lines, *short dashes*; result lines, *solid*. (See Fig. 55.)

(*b*) All lines to be in *color* and drawn *full*. Given lines, *blue*; construction lines, *red*; result lines, *black*.

The French curve should be used in inking the irregular curves.

Problem 1.— *To bisect a straight line or a circular arc.*

Let *AB* or *AE'B* be the given line. With any appropriate radius, centers *A* and *B*, describe arcs intersecting in points *C* and *D*. Draw *CD* intersecting *AB* and *AE'B* in *E* and *E'*, the required middle points.

Problem 2.— *To draw a line perpendicular to a given line at a given point in the line.*

Let *AB* be the given line, and *C* the given point. With any radius, *C* as center, draw arcs intersecting *AB* in *D* and *E*. With any appropriate radius, centers *D* and *E*, describe arcs intersecting in point *F*. Draw *FC*, the required perpendicular.

Problem 3.— *To draw a perpendicular to a line at or near its extremity. First Method.*

Let *AB* be the given line, and *A* the given point. With any appropriate radius, center *A*, draw the arc *CDE*. With the same radius, center *E*, intersect arc *CDE* in the point *D*. With the same radius, center *D*, intersect arc *CDE* in *C*. With the same or any appropriate radius, centers *C* and *D*, draw arcs intersecting in *F*. Draw *FA*, the required perpendicular.

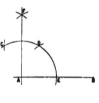

Problem 4.— *To draw a perpendicular to a line at or near its extremity. Second Method.*

Let *AB* be the given line, and *A* the given point. Assume any point *D*. With radius *AD*, center *D*, draw the arc *CAB*, intersecting *AB* at *B*. Draw from *B* through point *D*, to intersect arc *CAB* in *C*. Draw *AC*, the required perpendicular.

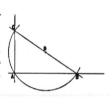

Problem 5.— *To draw a perpendicular to a line from a point outside the line.*

Let *AB* be the given line, and *C* the given point. With any appropriate radius, center *C*, draw an arc intersecting *AB* in points *D* and *E*. With any appropriate radius, centers *D* and *E*, draw arcs intersecting in point *F*. Draw *CF*, the required perpendicular.

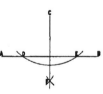

Problem 6.— *To draw a perpendicular to a given line from a point opposite to the end of the line.*

Let *AB* be the given line, and *C* the given point. Draw a line from *C* to any point, as *B*, in *AB*. Bisect *CB* at *D* (Prob. 1). With *CD* as radius, center *D*, draw arc *CAB*, intersecting the given line at *A*. Draw *CA*, the required perpendicular.

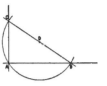

Problem 7.— *To draw a line at a given distance from and parallel to a given line.*

Let *AB* be the given line, and *CD* the given distance. With radius *CD*, any two assumed points *E* and *F* on the line as centers, draw arcs *GH* and *JK*. Erect perpendiculars to *AB* at points *E* and *F* (Prob. 3), intersecting arcs *GH* and *JK* in points *H* and *J*. Through points *H* and *J* draw *HJ*, the required line.

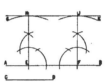

Problem 8.— *To draw a line parallel to a given line and passing through a given point.*

Let *AB* be the given line, and *D* the given point. With any appropriate radius, center *D*, draw arc *EC*. With the same radius, center *E*, describe arc *DF*. With chord *DF* as radius, center *E*, intersect arc *CE* in point *C*. Draw *CD*, the required line.

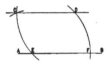

Problem 9.— *To bisect an angle.*

Let *BAC* be the given angle. With any appropriate radius, center *A*, describe an arc intersecting *AB* and *AC* in points *B* and *C*. With any radius, centers *B* and *C*, describe arcs intersecting in point *D*. Draw *AD*, the bisector of the given angle.

Problem 10.— *To trisect a right angle.*

Let *ABC* be the given right angle. With any appropriate radius, center *B*, describe an arc intersecting *AB* and *BC* in points *A* and *C*. With the same radius, centers *A* and *C*, intersect arc *AC* in points *E* and *D*. Draw *BD* and *BE*, the trisectors of the given right angle.

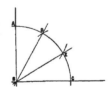

Problem 11.— *To construct an angle equal to a given angle.*

Let *CAB* be the given angle.   Draw *A'B'* of indefinite length.   With any equal radii, centers *A* and *A'*, draw arcs *CB* and *C'B'*.   With the chord *BC* as radius, center *B'*, intersect arc *B'C'* in *C'*.   Draw *A'C'* ; then angle *C'A'B'* is equal to the given angle *CAB*.

Problem 12.— *To divide a given line into any number of equal parts.   First Method.*

Let *AB* be the given line, and the required number of parts *five*.   Through point *A* draw *AC*, making any angle with *AB*.   Draw *BC'*, making angle *ABC'* equal to *CAB* (Prob. 11).   Take any distance as a unit, and lay it off on *AC* and *BC'* as many times as the required number of parts less one.   Draw lines *1–4'*, *2–3'*, *3–2'*, *4–1'*, dividing *AB* into the required number of equal parts.

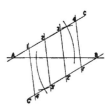

Problem 13.— *To divide a given line into any number of equal parts.   Second Method.*

Let *AB* be the given line, and the required number of parts *five*.   Draw *AC*, making any angle with *AB*. Lay off on *AC* any distance taken as a unit, as many times as the required number of parts.   Connect the last point of division, *5*, with point *B*.   Parallel to *5 B* draw lines through points *4, 3, 2, 1*, intersecting *AB* in points *4', 3', 2', 1'*, dividing *AB* into the required number of parts.

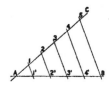

Problem 14.— *To divide a given line proportionally to a given divided line.   First Method.*

Let *AE*, divided into the parts *AB*, *BC*, *CD*, and *DE*, be given, and let *FG* be the line required to be divided proportionally to *AE*.   Draw *F'G'*, equal to *FG*, parallel to *AE*, and at any convenient distance from it. Draw *AF'* and *EG'* produced to intersect in point *H*. Draw *BH*, *CH*, and *DH*, intersecting *F'G'* in points *B'*, *C'*, and *D'*, marking the required divisions.

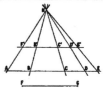

Problem 15.— *To divide a given line proportionally to a given divided line. Second Method.*

Let *AB*, divided by the points *E, F, G, H, J*, be given, and *CD* the line required to be divided proportionally to *AB*. Draw *AK* at any convenient angle with *AB*, and make *AD'* equal to *CD*. Draw *BD'*, and parallel to it draw *EE'*, *FF'*, etc., giving the points *E'*, *F'*, *G'*, *H'*, *J'*, which mark the required divisions.

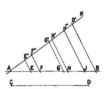

Problem 16.— *To find the distance which is the fourth proportional to three given distances.*

Let *AB*, *CD*, and *EF* be the given distances. Draw *GJ* of indefinite length, and lay off *GH* equal to *AB*, and *HJ* equal to *EF*. Draw *GL* of indefinite length, and making any convenient angle with *GJ*. On *GL* lay off *GK* equal to *CD*, and draw *KH*. Through *J*, and parallel to *KH*, draw *LJ* intersecting *GL* in point *L*. Distance *KL* is the required fourth proportional; that is, *AB* is to *CD* as *EF* is to *KL*.

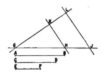

Problem 17.— *To find the distance which is the mean proportional between two given distances.*

Let *AB* and *CD* be the given distances. Draw *EH* of indefinite length, and lay off *EG* equal to *AB*, and *GH* equal to *CD*. Bisect *EH* (Prob. 1) in point *F*. With radius *EF*, center *F*, draw the semicircle *EJH*. At point *G* erect a perpendicular to *EH* (Prob. 2), intersecting the semicircle in point *J*. Distance *JG* is the required mean proportional; that is, *AB* is to *JG* as *JG* is to *CD*.

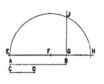

Problem 18.— *To draw a circle through three points not in the same straight line, or to circumscribe a circle about a triangle.*

Let *A, B,* and *D* be the given points or *ABD* the given triangle. Bisect *AB* and *AD* by lines *EG* and *FH*, intersecting in point *C*, the center of the required circle.

Note.— To find the center of a circle, assume any three points in its circumference and use the same construction.

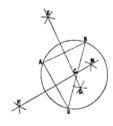

Problem 19.— *To draw a circular arc through three points not in the same straight line, when the center is not accessible.*

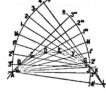

Let *A*, *E*, and *J* be the given points. Draw *AJ*. With *AJ* as radius, centers *A* and *J*, draw the arcs *AMP* and *RNJ*. Draw *AE* produced to intersect arc *RNJ* in point *N*, and *JE* produced to intersect arc *AMP* in *M*. Above and below points *M* and *N* lay off on arcs *AMP* and *RNJ*, with any convenient unit, equal spaces as *M 1*, *M 1″*, *N 1′*, *N 1‴*, etc. Draw *J 1* and *A 1′* intersecting in *F*, a point in the required arc. Draw *J 2* and *A 2′* intersecting in *G*, a second point in the required arc. Locate in like manner, aided by inspection of the figure, *B*, *C*, *D*, *K*, *L*, the rest of the points determining the required arc.

Problem 20.— *Through a given point to draw a line which shall pass through the inaccessible intersection of two given lines.*

Let *AB* and *CD* be the given lines, *P* the given point. Assume any two points *E* and *E′* on *AB*, and any point *F* on *CD;* draw *PE*, *PF*, and *EF*. Through *E′* draw *E′F′* parallel to *EF;* draw *E′P′* parallel to *EP*, and *F′P′* parallel to *FP*, intersecting in point *P′*. Draw *PP′*, the required line.

Problem 21.— *To construct a triangle, the lengths of its three sides being given.*

Let *AB*, *CD*, and *EF* be the given sides, and *AB* be the base. With *CD* as radius, center *A*, describe an arc at *G*. With *EF* as radius, center *B*, intersect the preceding arc in point *G*. Draw *AG* and *GB*, completing the required triangle *AGB*.

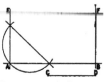

Problem 22.— *To construct a rectangle, the lengths of its sides being given.*

Let *AB* and *CD* be the given sides. At either end of *AB*, as *A*, draw a perpendicular (Prob. 4) of indefinite length, and upon it lay off *EA* equal to *CD*. With *AB* as radius, center *E*, describe an arc at *F*. With *CD* as radius, center *B*, intersect the preceding arc in point *F*. Draw *EF* and *FB*, completing the required rectangle.

Problem 23.— *To construct a polygon equal to a given irregular polygon.*

Let *ABCDFE* be the given polygon. Draw lines dividing the given polygon into triangles, as *ABE*, *BEC*, etc. Draw *E'F'* equal to *EF*, and on *E'F'* construct triangle *C'E'F'*, equal to triangle *CEF* (Prob. 21). On *E'C'* construct triangle *E'B'C'*, equal to triangle *EBC*. By similar construction draw the triangles *E'A'B'* and *C'D'F'*, completing the required polygon.

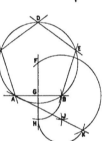

Problem 24.— *To construct a regular pentagon, the length of one side being given.*

Let *AB* be the given side. Bisect *AB* by the perpendicular *FH*. Make *GH* equal to *GB*. With radius *GB*, centers *B* and *H*, draw arcs intersecting at *J*. Draw *AJ* produced, and make *JK* equal to *GB*. With radius *KB*, center *B*, intersect *FH* at *F;* with the same radius, center *F*, draw the circle *ABD*. With *AB* as radius, start at *A*, and cut the circle in points *C*, *D*, and *E*. Draw *AC*, *CD*, *DE*, and *EB*, completing the required pentagon.

Problem 25.— *To construct a regular hexagon, the length of one side being given.*

Let *AB* be the given side. With radius *AB*, centers *A* and *B*, describe arcs intersecting in point *C*. With radius *AB*, center *C*, draw the circle *AEB*. With the same radius, starting at point *A*, cut the circle in points *D*, *E*, *F*, *G*, locating the remaining sides of the required hexagon.

Problem 26.— *To construct a regular pentagon, the circumscribing circle being given.*

Let *AEB* be the given circle. Draw any diameter, *AB*, of the circle, and at its center, *C*, draw *EC* perpendicular to *AB*. Bisect *CB* in *K*, and with *EK* as radius, center *K*, draw arc *EJ*. With chord *JE* as radius, center *E*, intersect the given circle in *D*. *DE* is one side of the required pentagon.

Problem 27.— *To construct a regular polygon of any number of sides, the length of one side being given.*

Let *AB* be the given side, and the number of sides *seven*. With radius *AB*, *A* as center, draw the semicircle *1-4-B*. Divide the semicircle into as many equal parts as there are sides in the required polygon. Draw *AD*, which is a side of the required polygon, connecting point *A* with the *second* point of division in the semicircle. Bisect *AB* and *AD* (Prob. 1), and produce the

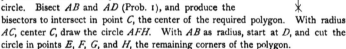

bisectors to intersect in point *C*, the center of the required polygon. With radius *AC*, center *C*, draw the circle *AFH*. With *AB* as radius, start at *D*, and cut the circle in points *E*, *F*, *G*, and *H*, the remaining corners of the polygon.

Problem 28.— *To find the point of tangency of a straight line and a circle.*

Let *AB* be the given straight line, and *DF* the given circle, described from *C*. From *C* draw *CE* perpendicular to *AB* (Prob. 6), intersecting *AB* at *E*, the required point of tangency.

Problem 29.— *To draw a circular arc tangent to a straight line and to a circle at a given point.*

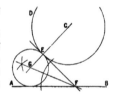

Let *AB* be the given straight line and *E* the given point on circle *DE* described from point *C*. Draw *CE* produced. Draw *EF* perpendicular to *CE* (Prob. 2). Bisect angle *EFA* (Prob. 9). Point *G*, the intersection of the bisector and of *CE* produced, is the center of the required arc.

Problem 30.— *To connect two given lines by a reversed curve, given one point of tangency and the radii of the two curves.*

Let *AB* and *CD* be the given lines, and *B* the given point of tangency. At *B* draw the indefinite line *EF* perpendicular to *AB* (Prob. 4). Make *BE* equal to one given radius, and *BF* equal to the other given radius. Draw the indefinite line *HJ* parallel to *CD* and at a distance equal to *EB* from *CD* (Prob. 7). With radius *EF*, center *F*, draw arc *EJ* to intersect *HJ* in point *J*. From *J* draw *JC* perpendicular to *CD*, intersecting *CD* at *C*. With radius *BF*, center *F*, draw arc *BK*. With radius *CJ*, center *J*, draw arc *KC*. *BKC* is the required curve.

Problem 31.— *To connect two given parallel lines by a reversed curve, given one point of tangency and the point of reversed curvature.*

Let *AB* and *CD* be the given lines, *A* the given point of tangency, and *K* the given point of reversed curvature. Draw *AK* produced to meet *CD* at *D*. At points *A* and *D*, draw the indefinite lines *AE* and *DF*, perpendicular to *AB* and *CD*. Bisect *AK* and produce the bisector to meet *AE* at *E*. Bisect *KD* and produce the bisector to meet *DF* at *F*. With radius *AE*, center *E*, draw arc *AK*. With radius *DF*, center *F*, draw arc *KD*. *AKD* is the required curve.

Problem 32.— *To connect two given parallel lines by a reversed curve, given the points of tangency and the ratio of the radii.*

Let *AB* and *CD* be the given parallel lines, *B* and *C* the given points of tangency, and let the required radii be in the ratio of *HJ* to *JK*. At the points of tangency *B* and *C*, draw the indefinite lines *BE* and *CF*, perpendicular to the given lines. Connect *B* and *C* and find point *G*, so that *BG* is to *GC* as *HJ* is to *JK* (Prob. 15). Bisect *BG* and produce the bisector to meet *BE* at *E*. Bisect *GC* and produce the bisector to meet *CF* at *F*. With radius *BE*, center *E*, draw arc *BG*. With radius *CF*, center *F*, draw arc *CG*. *BGC* is the required curve.

Problem 33.— *To connect two given non-parallel lines by a reversed curve, given the points of tangency and the ratio of the radii.*

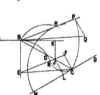

Let *AB* and *CD* be the given lines, *B* and *C* the given points of tangency. At points *B* and *C*, draw the indefinite lines *BE* and *CF*, perpendicular to *AB* and *CD*. Connect *B* and *C*. Bisect *BC* at *G* and with radius *BG*, center *G*, draw the indefinite arc *BH*. On *BC* find point *J*, so that the ratio of *BJ* to *JC* is the given ratio of the radii (Prob. 15). From *J* draw *JK* perpendicular to *AB*, and *JL* perpendicular to *CD*. Make *LM* equal to *BK*. With radius *MC*, center *C*, describe an arc intersecting arc *BH* in *H*. Draw *BH* of indefinite length, and make *BP* equal to *BC*. From *P* draw *PQ* perpendicular to *CD*, make *PQ* equal to *CJ*, and draw *BQ*. From *C* draw *CE* parallel to *BQ*, intersecting *BE* in *E*. From *E* draw *EF* parallel to *BP*, intersecting *CF* at *F*. With radius *BE*, center *E*, draw arc *BN*. With radius *CF*, center *F*, draw arc *NC*. *BNC* is the required curve.

Problem 34.—*To draw a circle tangent to two given circles and at a given point in one of them* (*two solutions*). *First Method.*

Let *ABD* and *EFH* be the given circles, and *B* the given point. Draw from point *B* through *C*, the center of circle *ABD*, and produce the line indefinitely.

First solution. Make *BJ* equal to the radius of circle *EFH*. Draw *JG*, bisect it, and produce the bisector to intersect *CB* produced in point *L*. With *BL* as radius, center *L*, draw arc *BO* of the required circle.

Second solution. Make *BM* equal to the radius of circle *EFH*. Draw *MG*, bisect it, and produce the bisector to intersect *CL* in *N*. With *BN* as radius, center *N*, draw *BP*, the required circle.

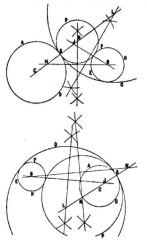

Problem 35.— *To draw a circle tangent to two given circles and at a given point in one of them* (*two solutions*). *Second Method.*

Let *ABD* and *MEJ* be the given circles, and *B* the given point. Draw from point *B* through *C*, the center of circle *ABD*, and produce the line indefinitely. Through *G*, the center of circle *MEJ*, draw a line parallel to *BC*, cutting the circle *MEJ* in points *E* and *F*.

First solution. Draw *BE* produced to meet the circle *MEJ* in point *J*. Draw *JG* produced to meet *BC* at *L*. With *BL* as radius, center *L*, draw arc *BOJ* of the required circle.

Second solution. Draw *BF* intersecting circle *MEJ* in point *M*. Draw *GM* produced to meet *BC* at *N*. With *BN* as radius, center *N*, draw *BMP*, the required circle.

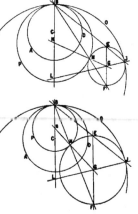

Problem 36.— *To inscribe a circle within a triangle.*

Let ABD be the given triangle. Bisect any two angles of the triangle (Prob. 9). The intersection C of the bisectors is the center of the required circle.

Note.— If adjacent exterior angles, as EAB and ABF, be bisected, the bisectors will intersect in a point, C', which is the center of a circular arc tangent to one side of the triangle and to two of its sides produced.

Problem 37.— *Within an equilateral triangle to inscribe three equal circles, each tangent to the others and to two sides of the triangle.*

Let ABC be the given equilateral triangle. Bisect the sides in points D, E, and F. Draw FA, DC, and BE. With radius DE, centers D, E, and F, draw arcs EF, FD, and DE. The intersections G, J, and H are the centers of the required circles.

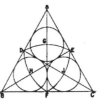

Problem 38.— *Within an equilateral triangle to inscribe six equal circles, tangent to each other.*

Let ABC be the given equilateral triangle. Bisect the sides in points D, E, and F. Draw BE, DC, and FA. Bisect the angle EBC by line BG, intersecting AF in point G. Make DJ and EK each equal to FG, and through points J, K, and G, draw LM, LH, and HM, parallel to the sides BC, BA, and AC of the triangle. With GF as radius, centers L, J, H, K, M, and G, describe the required circles.

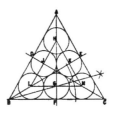

Problem 39.— *Within a given circle to inscribe three equal circles tangent to each other.*

Let BDG be the given circle, center C. Divide the circumference into six equal parts by making the chords AD, DF, etc., each equal to the radius CA of the circle. Draw the diameters AG, BF, and ED. Produce any diameter, as BF, and make FH equal to CF. Draw GH and bisect the angle CHG by the line HJ, intersecting CG in point J. With radius JC, center C, draw the circle MKL, intersecting the diameters in points M, K, and L, the centers of the required circles.

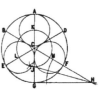

Problem 40.— *Within a given circle to inscribe any given number of equal circles tangent to each other.*

Let *DAE* be the given circle, and the required number of inscribed circles *five*. Divide the circle *DAE* into ten equal parts (Prob. 26), and draw the diameters *AB*, *DE*, etc. At the extremity of any diameter, as *B*, draw the tangent *FG* perpendicular to *AB*, and produce the adjacent diameters to intersect the tangent at *F* and *G*. In the triangle *CFG* find the center *H* of the inscribed circle (Prob. 36). With radius *HC*, center *C*, draw the circle *LJK*, giving the remaining required centers, *L*, *J*, *K*, and *M*.

Problem 41.— *To draw an ellipse, given the rectangular axes. First Method.*

Let *AB* be the major axis, and *DE* the minor axis. With radius equal to one-half the major axis, centers *D* and *E*, intersect *AB* in points *F* and *F'*, the *foci* of the

required curve. Divide by eye, *FC* and *CF'* into any number of parts, decreasing from *C* to *F* and from *C* to *F'*. To locate the points of the curve: With *A5* as radius, center *F*, describe an arc at *L*. With *B5* as radius, center *F'*, intersect the preceding arc in *L*, a point in the required curve. With radii *A4*, *A3*, *A2*, and *A1*, center *F*, describe arcs at *K*, *J*, *H*, and *G*. With radii *B4*, *B3*, *B2*, and *B1*, center *F'*, intersect in the same order the arcs at *K*, *J*, *H*, and *G*, four additional points in the required curve. Repeat in each of the remaining quadrants.

Problem 42.— *To draw a tangent to an ellipse at a given point in the curve. First Method.*

Let *Q* be the given point. Find the foci *F* and *F'* (Prob. 41). Draw *F'Q* and *FQ* produced. Bisect angle *F'QR* (Prob. 9). The bisector *ST* is the required tangent.

Problem 43.— *To draw a tangent to an ellipse from a given point outside the curve. First Method.*

Let *M* be the given point. Find the foci *F* and *F'* (Prob. 41). With *FM* as radius, center *M*, draw arc *FN*. With *AB* as radius, center *F'*, intersect arc *FN* in point *N*. Draw *NF'*. Bisect the angle *FMN* (Prob. 9) and draw the bisector, intersecting *F'N* at *P*. *MP* is tangent to the ellipse at point *P*.

Problem 44.—*To draw an ellipse, given the rectangular axes. Second Method.*

Let *AB* and *DE* be the axes. With *AC* and *DC* as radii, center *C*, describe the major and minor auxiliary circles *AGQ* and *DJ'E*. Assume points in the outer circle, as *G, H, J*. Draw radii *CG, CH*, and *CJ*. Draw lines parallel to *AB* from points *G'*, *H'*, and *J'*, intersected by lines parallel to *DC* from *G, H*, and *J* in *G'', H''*, and *J''*, points in the required curve. Repeat in each of the remaining quadrants.

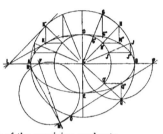

Problem 45.— *To draw a tangent to an ellipse at a given point in the curve. Second Method.*

Let *K* be the given point on the ellipse. At *K'*, the corresponding point on the major auxiliary circle (see Prob. 44), draw the tangent *LK'*, perpendicular to the radius *K'C*, meeting the major axis at *L*. Draw *LK*, the required tangent.

Note.— The minor auxiliary circle and minor axis may be similarly used.

Problem 46.— *To draw a tangent to an ellipse from a given point outside the curve. Second Method.*

Let *M* be the given point. Find either focus, as *F* (Prob. 41). On *MF* as diameter, draw a circle intersecting the major auxiliary circle in points *N* and *O*, through which draw *MN* and *MO*, the required tangents.

To find the exact point of tangency, as of *MP*. Take the point *P* where the tangent *MO* intersects the major axis. On *CP* as diameter draw a semicircle, intersecting the major auxiliary circle at *Q*. Find *Q''*, the point on the ellipse corresponding to *Q* (Prob. 44). Point *Q''* is the point of tangency.

Note.— The minor axis and minor auxiliary circle may be similarly used.

Problem 47.—*To draw an ellipse, given the oblique axes.*

Let *AB* be the major axis, and *DE* the minor axis. Through point *E* draw *HJ* parallel to *AB*, and through point *A*,

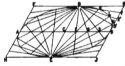

HF parallel to *ED;* complete the parallelogram *HFGJ*, containing the points *D* and *B*. Divide *CB* and *GB* into any number of similarly spaced parts, as shown by *1, 2, 3*, and *1', 2', 3'*. Draw *D1', D2', D3'*, and *E1, E2, E3* produced. The intersections *K, L*, and *M*, are points in the required curve.

Note.— The same construction applies when the axes are at right angles.

Problem 48.— *To find the axes of symmetry (rectangular axes) of an ellipse.*

Let AC and CB be the oblique semi-axes. Draw EG parallel to CB. Draw AD perpendicular to EG and make AD equal to CB. Draw CD, bisect it, and produce the bisector to intersect AG in F. With DF as radius, center F, draw the circle EDG, intersecting EG in E and G. Draw from G through C, and from E through C, giving LM and NP, the major and minor axes respectively.

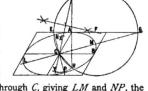

To find the extremities of the rectangular axes. Produce DA to intersect circle DGH in point H. Connect C and H. Make CX and CY each equal to CH. Bisect DY and lay off CL and CM, each equal to one-half of DY; bisect DX and lay off CN and CP, each equal to one-half of DX. Points L, M, N, and P are the extremities of the rectangular axes.

Problem 49.— *To draw a parabola, given the focus and the directrix.*

Let AR' be the directrix of the parabola, BC its axis, and F the focus. Bisect BF in point D, the vertex of the parabola. Assume on the axis BC, any points, as G, H, ... M, and through these points draw the indefinite lines $G'G''$, $H'H''$, etc., parallel to AR'. With BG as radius, center F, intersect $G'G''$ in points G' and G''; with radius BH, center F, intersect $H'H''$ in H' and H'', and so on.

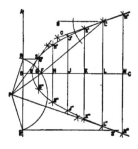

Problem 50.— *To draw a tangent at a given point on a parabola. First Method.*

Let L' be the given point. Draw from L' to the focus F. Through L' draw $L'N$ parallel to BC. Bisect the angle $NL'F$ by $L'O$, the required tangent.

Problem 51.— *To draw a tangent to a parabola from a given point outside the curve. First Method.*

Let P be the given point. Connect P with the focus F. With FP as radius, center P, draw an arc intersecting the directrix AR' in points R and R'. Bisect the angles RPF and FPR'; the bisectors PS and PS' are the required tangents.

To find the exact points of tangency. From R and R' draw lines parallel to BC, intersecting the tangents PS and PS' in S and S', the required points.

Problem 52.—*To draw a parabola, given the axis, the vertex, and a point on the curve.*

Let AB be the axis of the parabola, A its vertex, and D a point on the curve. Complete the rectangle $ABDC$. Divide AC into any number of equal parts, say *five*, by points $1, 2, 3, 4$. Divide CD into the same number of equal parts by points $1', 2', 3', 4'$. Through points $1, 2, 3, 4$, draw lines parallel to AB. Draw $A1', A2',$ $A3', A4'$, intersecting the parallels from

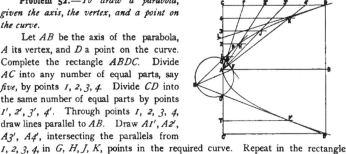

$1, 2, 3, 4$, in G, H, J, K, points in the required curve. Repeat in the rectangle $ABD'C'$.

To find the focus. From point 1 draw a line perpendicular to $A4'$ (Prob. 5), produced to intersect the axis AB in F, the required point.

Problem 53.— *To draw a tangent at a given point on a parabola. Second Method.*

Let D be the given point. Draw CD parallel to AB, intersecting AC in point C. Bisect AC in point L. Draw LD, the required tangent.

Problem 54.— *To draw a tangent to a parabola from a given point outside the curve. Second Method.*

Let N be the given point. Connect N with the focus F (Prob. 52). On NF as diameter draw a circle intersecting CC' in points O and O'. Through O and O' draw NP and NP', the required tangents.

To find the exact point of tangency. Make $O'T$ equal to $O'A$. Through T draw TP' parallel to AB, intersecting NP' in P', the required point of tangency.

Problem 55.—*To draw a hyperbola, given the major axis and one point on the curve.*

Let AB be the major axis, and D the given point. Draw the rectangle $BCDE$. Divide CD and ED each into any number of equal parts, in this case *four*. Draw $B1,$ $B2, B3,$ and $A1', A2', A3'$, intersecting in

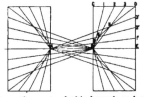

$F, G,$ and H, points in the required curve. Draw the rest of this branch and the opposite branch of the curve by the same construction.

Problem 56.—*To find the asymptotes and foci of a hyperbola.*

Let AB be the major axis of the hyperbola, and G a known point on the curve. Bisect AB by the perpendicular DH (Prob. 1). With radius AC, center C, draw the circle ABH.

To find the asymptotes. At H draw the tangent HJ parallel to AB. Through the known point of the curve, G, draw DG parallel to AB. With DG as radius, center C, draw an arc intersecting HJ in point J. Make DE equal to HJ and draw CE, one required asymptote. Make angle KCL equal to angle LCM (Prob. 11), and draw KC, the other required asymptote.

To find the foci. At the points of intersection, N and N', of ME with the circle ABH, draw NF and $N'F'$ perpendicular to ME, intersecting the major axis AB in points F and F', the required foci.

Note.—The circle ABH is known as the major auxiliary circle.

Problem 57.—*To draw a tangent to a hyperbola at a given point in the curve.*

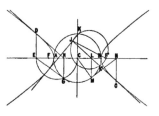

Let D be the given point. Draw the major auxiliary circle on AB as diameter. From D draw DE perpendicular to AB (Prob. 5). On EC as diameter, draw the semicircle EGC, intersecting the major auxiliary circle in point G. From G draw GH perpendicular to AB. Draw DH, the required tangent.

Problem 58.—*To draw a tangent to a hyperbola from a given point outside the curve.*

Let J be the given point. Connect J with either focus, as F'. On JF' as diameter, draw a circle intersecting the major auxiliary circle at points K and K'. Draw KJ and JK', the required tangents.

To find the exact point of tangency. Tangent JK' intersects the major axis AB in point L. At L draw LM perpendicular to AB and intersecting the major auxiliary circle at M. Draw CM, and perpendicular to CM draw MN. Draw NO perpendicular to AB, intersecting the tangent JK' in point O, the required point of tangency.

Problem 59.— *To draw an ellipse with a trammel, given the rectangular axes.*

Let AB and CD be the given axes. On the edge of a strip of paper or cardboard, mark the distance $A'E'$ equal to the semi-major axis AE. Mark the distance $E'F$ equal to the semi-minor axis CE. Place the trammel so that point A' falls on the axis CD, produced if necessary, and point F on the axis AB; then point E' will be a point of the required ellipse. Find as many points as desired, and draw the curve.

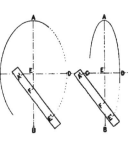

Problem 60.— *To draw a curve approximating an ellipse, composed of circular arcs, given the major axis.*

Let AB be the given major axis. Make AC and $C'B$ each equal to five-sixteenths of AB (Prob. 15). With AC as radius, centers C and C', draw the circles ADE' and DBF, intersecting in D and D'. Draw DC, DC', $D'C$, and $D'C'$, all produced. With radius $E'D$, centers D and D', draw arcs $E'F'$ and EF, completing the required curve.

Note.— The minor axis of this curve is three-quarters of the major axis.

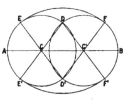

Problem 61.— *To draw a curve approximating an ellipse, composed of circular arcs, given the major and minor axes. First Method.*

Let AB be the major axis and DE the minor axis. Draw DB. Make CG equal to CD and make DH equal to GB. Bisect HB by line $J'K$, intersecting CB in L', and CE produced in J'. Make CJ equal to CJ', and CL equal to CL'. Draw $J'L$, JL', and JL, all produced. To draw the curve: With radius $J'D$, centers J' and J, draw the arcs MDN and $M'EN'$. With radius LA, centers L and L', draw the arcs MAM' and NBN'.

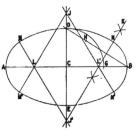

Problem 62.— *To draw a curve approximating an ellipse, composed of circular arcs, given the major and minor axes. Second Method.*

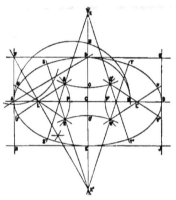

Let *AB* be the major axis, and *DE* the minor axis. Produce *DE* indefinitely in both directions. Through *E*, draw *HJ* parallel to *AB*, and complete the rectangle *HFGJ*, its sides passing through points *A*, *D*, and *B*. Draw *AD*. From point *F* draw a perpendicular to *AD* (Prob. 5), intersecting *AB* in *L* and produced to intersect *DE* produced in *K'*. Make *CK* equal to *CK'*, and *CL'* equal to *CL*. With radius *DC*, center *C*, draw an arc intersecting *AB* in point *N*. On *AN* as a diameter draw the semicircle *AMN*. Make *CO* and *CO'* each equal to *MD*, and with radius *OK*, centers *K* and *K'*, draw arcs through points *O* and *O'*. Make *AP* and *P'B* each equal to *MC*, and with radius *LP*, centers *L* and *L'*, intersect the arcs drawn through *O* and *O'* in points *Q*, *Q'*, *R*, and *R'*. Draw, of indefinite length, *LQ, LR, L'Q'*, and *L'R'*. Draw, of indefinite length, *K'R, K'R', KQ*, and *KQ'*. To draw the required curve: With radius *DK'*, centers *K'* and *K*, describe arcs *SDT* and *S'ET'*. With radius *SR*, centers *R, R', Q*, and *Q'*, draw arcs *SU, S'U', TV*, and *T'V'*. With radius *AL*, centers *L* and *L'*, draw arcs *UAU'* and *VBV'*.

Problem 63.— *To draw an oval, given its width.*

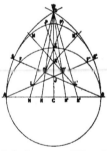

Let *AB* be the width of the oval. On *AB* as diameter draw the circle *AEB*. With radius *AB*, centers *A* and *B*, draw arcs intersecting in *D*. Draw *DC*. Bisect arcs *AE* and *EB* in points *F* and *F'*. Draw *AF'* and *BF* produced to intersect arcs *AGD* and *DG'B* in *G* and *G'*. Bisect *AC* and *CB* in *H* and *H'*. Draw *HD* and *H'D*, intersecting *AG'* and *GB* in *L* and *L'*. Bisect *EJ* in *K* and draw *LK* and *L'K* produced. With radius *GL'*, centers *L'* and *L*, draw arcs *GM* and *M'G'*. Bisect *HC* and *CH'* in *N* and *N'*. Draw *ND* and *N'D*, intersecting *LM'* and *ML'* in *O* and *O'*. Draw *OE* and *O'E* produced. With radius *MO'*, centers *O'* and *O*, draw arcs *MP* and *P'M'*. With radius *EP*, center *E*, draw arc *PP'*, completing the required curve.

Problem 64.— *Given the short diameter, to draw an oval the long diameter of which shall be 1½ times the short diameter.*

Let *AB* represent the short diameter. Bisect *AB* and draw the perpendicular *DE* produced. With center *C*, draw the circle *ADE*. Produce *AB* in each direction, and make *AF*, *BF'*, and *EG*, each equal to *AC*. Bisect *EG* in *H*. Draw *FH* and *F'H* produced. With radius *FB*, centers *F* and *F'*, draw arcs *BJ'* and *AJ*. With radius *GH*, center *H*, draw arc *JGJ'*, completing the required curve *ADBJ'GJ*.

Problem 65.— *To draw a variable spiral.*

Let *CD* be the measure of the required curve. Divide *CD* into eight equal parts, as indicated by the points *1*, *2*, ... *7*. Upon division *4–5* as diameter draw a circle, the eye of the spiral. (See also the enlargement, Fig. *B*.) On *4–5* as a diagonal draw the square *E4F5*, its diameters *GK* and *JH*, and the square *GHKJ*. Divide *GL* into *two* equal parts, and through the point of division draw *5'4'*. Divide *LM* into *six* equal parts, and through the points of division draw *8'7'*, *4'3'*, *1'2'*, and *5'6'*. Divide *MK* into *three* equal parts, and through the points of division draw *3'2'* and *7'6'*. To draw the curve: With *4–1'* as radius, center *1'*, draw the arc *4N*, carried to the line *2'1'* produced. With radius *N2'*, center *2'*, draw the arc from point *N* to *O*, lying in *3'2'* produced. With radius *O3'*, center *3'*, carry the arc to *4'3'* produced, and so on. Take for successive centers the points *4'*, *5'*, *6'*, *7'*, *8'*, *J*, *K*, *H*, and *G*, and draw the arc described from each center to the line drawn through the center used and the next one in advance.

Problem 66.— *To draw an Ionic volute.*

Let *AB*, Fig. *A*, be the measure of the required curve. Divide *AB* into seven equal parts. Through the fourth point of division from *A* draw *CD*, indefinite in length and perpendicular to *AB*. Take for the center of the spiral any point on *CD*, and from this point, with radius equal to one-half of a division of *AB*, describe circle *FGHJ*, the eye of the volute. (See also its enlargement, Fig. *B*.) Draw the diameter of the eye, *GJ*, perpendicular to *FH*. On *FH* as a diagonal draw the square *FGHJ*, its diameters *KN* and *ML*, and the square *KLNM*. Divide *KM* into six equal parts. Draw *1″2″* parallel to and distant from *FH* one-half of *3′4′*. Draw *9″10″* parallel to and distant from *MN* one-half of *5′M*. Project points *1′* and *2′* to *GO*, and through the points thus obtained draw *8″7″* and *4″3″* parallel to *KL*. Draw *5″6″* parallel to and midway between *1″2″* and *9″10″*. Draw *8″9″*, *4′5″*, *3″2″*, and *7″6″* parallel to

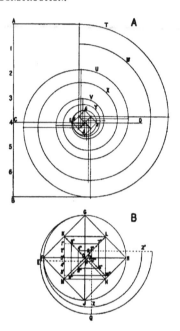

KM and separated by spaces each equal to a division of *KM*. *To draw the outer spiral TUV* (Fig. *A*): Take *G1″* as radius, center *1″*, and draw arc *GE* tangent to the eye of the volute at *G*, and carried to the line *2″1″* produced. With radius *E2″*, center *2″*, draw from *E* to *Q*, lying in *3″2″* produced. With radius *Q3″*, center *3″*, describe an arc from *Q* to *4″3″* produced, and so on. Take for successive centers the points *4′5″6″7″8″9″10″*, *L*, and *K*, and draw the arc described from each center to a line drawn through the center used and the next one in advance. *To draw the inner spiral WXY* (Fig. *A*): Locate the point *P* in *OH*, midway between point *O* and line *3″2″*. Locate the point *S* midway between *P* and *2″*. With radius *FS*, center *S*, draw an arc from *F* to *Z*, lying in *3″2″* produced. Locate point *R*, making, by eye, distance *3″R* equal to *2″S*. With radius *ZR*, center *R*, draw an arc from *Z* to *Z′*, lying in *4″3″* produced. The dot placed between *4″* and *O* indicates the next center, and is placed a distance from *4″* equal to *3″ R*. Continue in like manner

Problem 67.— *To draw an Archimedean spiral.*

Let *AN* be the measure of the spiral. With *NA* as radius, center *A*, describe the circle *9′N3′*. Divide line *AN* and the circle *9′N3′* into the same number of equal parts, as *twelve.* Draw *A1′*, *A2′*, . . . *A11′*. With *A* as center, draw arcs from each numbered point in *AN*, to intersect the correspondingly numbered line drawn from center *A* to the circle *9′N3′*, giving *A*, *B*, *C*, . . . *N*, which are points in the required curve.

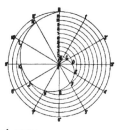

Note.— To obtain a spiral of two turns, divide *AN* into twice as many parts as the circle *9′N3′*.

Problem 68.— *To draw the involute of a circle.*

Let *A3H* be the given circle. Lay off on the circle equal spaces as indicated by the points *1*, *2*, . . . *5*, and at each point draw a tangent to the circle. Lay off the chord *1A*, once on tangent *1B*, twice on *2C*, three times on *3D*, etc., giving *A*, *B*, *C*, . . . *G*, points in the required curve.

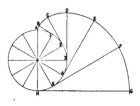

Note.— In this and the succeeding problems a more accurate result is obtained by using the true length of the circular arc instead of its chord (Prob. 73).

Problem 69.— *To draw a cycloid.*

Let *A*, lying in circle *BDA* which rolls on the straight line *AE*, be the generatrix. Lay off any equal distances on circle *BDA* and line *AE*, as indicated by the points *1*, *2*,...*6*, and *1′*, *2′*,...*6′*. Through the center of the rolling circle draw *CC*vi parallel to *AE*. Find the consecutive positions of the center of the rolling circle by erecting perpendiculars

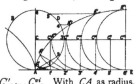

from *1′*, *2′*,...*6′*, to intersect *CC*vi in points *C′*, *C′′*,...*C*vi. With *CA* as radius, centers *C′*, *C′′*,...*C*vi, draw arcs tangent to *AE* at points *1′*,*2′*,...*6′*. To find points, as *A′*, *A′′*,...*A*ri, in the required curve. *First method.* Make the chord *1′A′* equal to the chord *A1*; make chord *2′A′′* equal to chord *A2*; make chord *3′A′′′* equal to chord *A3*. *Second method.* Intersect arcs *A′1′*, *A′′2′*, . . . *A*vi *6′*, by lines drawn parallel to *AE*, and passing through points *1*, *2*,...*6*.

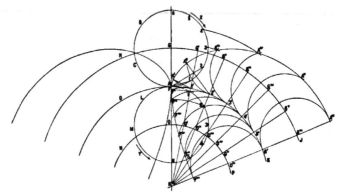

Problem 70.— *To draw an exterior epicycloid.*

Let point A, lying in circle BCA which rolls on circle DAE, be the generatrix. Starting at A, lay off any equal distances on circles BCA and DAE, as indicated by the points $1, 2, 3, \ldots 6$, and $1', 2', \ldots 6'$. Draw from the center K of the directing circle, through points $1', 2', \ldots 6'$, and produce the lines to intersect arc HGJ in points $G', G'', \ldots G^{vi}$, which are consecutive positions of the center G of the rolling circle. With radius GA, centers $G', G'', \ldots G^{vi}$, draw arcs tangent to the directing circle at points $1', 2', \ldots 6'$. To find points $A', A'', \ldots A^{vi}$ in the required curve. *First method.* Lay off on arc $1'A'$ the chord $1'A'$ equal to the chord $A1$; make chord $2'A''$ equal to chord $A2$; etc. *Second method.* With K as center, draw arcs, as $3A'''$ and $4A^{v}$, from the points of division in the rolling circle BCA to intersect, as at A''' and A^{iv}, each of its successive positions.

Problem 71.— *To draw a hypocycloid.*

Let point F, lying in circle MLF which rolls on circle DAE, be the generatrix. Starting at F, lay off any equal distances on MLF and DAE, as indicated by points $1, 2, \ldots 6$, and $1', 2', \ldots 6'$. Draw from the center K to $1', 2', \ldots 6'$, intersecting arc NOP in points $O', O'', \ldots O^{vi}$, which are consecutive positions of the center of the rolling circle. With FO as radius, centers $O', O'', \ldots O^{vi}$, describe arcs tangent to the directing circle at $1', 2', \ldots 6'$. To find points, as $F', F'', \ldots F^{vi}$, in the required curve. *First method.* Lay off on arc $1'F'$ the chord $1'F'$ equal to the chord $F1$; make chord $2'F''$ equal to chord $F2$; etc. *Second method.* With K as center, draw arcs, as $F'1$ and $F''2$, from the points of division in the rolling circle MLF to intersect, as at F' and F'', each of its successive positions.

Problem 72.— *To draw an interior epicycloid.*

Let point *A*, lying in circle *BAC* which rolls on circle *ADE*, be the generatrix. From *F*, the center of the directing circle, draw a circle *HJG*, passing through the center *G* of the rolling circle. Starting at point *A*, lay off on the directing circle any equal distances, as indicated by the points *1, 2,...6*. Draw *1F, 2F,...5F*, produced to intersect the path of the center

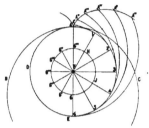

of the rolling circle in points *G', G'',... Gⁱ*. With each of these points as a center, radius *AG*, draw arcs tangent to the directing circle at *1, 2,...6*. From the tangent point of each arc, lay off the chord of arc *A1* once on arc *1A'*, twice on arc *2A''*, three times on *3A'''*, etc., giving *A', A'',... Aⁱ*, points in the required curve.

Problem 73.— *To lay off on a straight line the length of a given circular arc. An approximate method.*

Let *ABC* be the given arc. Draw *AD* tangent to arc *ABC*. Draw chord *CA* produced. Bisect *AC* at *E* and lay off *AF* equal to *AE*. With *CF* as radius, center *F*, intersect *AD* in point *G*; then *AG* is equal to arc *ABC*, very nearly.

Note.— If the arc subtends more than about 60°, it should be subdivided before using this construction, which becomes more accurate the smaller the subtended angle. Thus, if the arc subtends 60°, the length obtained is about one part in 900 too short; if 30°, about one part in 15,000.

Problem 74.— *To lay off on a circular arc of given radius the length of a given straight line. An approximate method.*

Let *AB* be the given straight line. With the given radius, draw arc *ACD* tangent to *AB*. Make *AE* equal to ¼ *AB*. With radius *BE*, center *E*, intersect arc *ACD* in point *F*; then arc *ACF* is equal to *AB*, very nearly.

Note.—The result is more accurate the smaller the angle subtended by the arc, substantially as in Prob. 73.

CHAPTER VI.

SELECTION AND ARRANGEMENT.

48. The Layout. This is a freehand or instrumental sketch in which is planned the general scheme of a proposed drawing. The principal requirements in making a layout are taste and good judgment in the selection and arrangement of the matter to appear in the drawing.

While in office practice selection and arrangement are likely to be restricted, more or less, on account of the fixed character of such work, yet, when opportunity offers, especially in architectural practice, taste and good judgment may be of decided value.* As these are personal qualities, it is evident that no hard-and-fast rules can be given for the guidance of the beginner; he must depend largely upon observation, criticism, and experience.

In order to emphasize the difference between good and bad selection and arrangement, first take a case illustrating the latter, as follows : Let it be required to select eight or ten geometrical constructions from the preceding chapter, and to arrange them on a sheet of some given size.

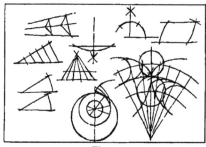

Fig. 77.

It is evident that the selection shown in Fig. 77 is bad, because the problems are not all of the same general character, the last two problems being wholly out of keeping with the remaining simple ones. As to the arrangement, the term does not apply to such haphazard placing — there is no semblance here of either order or taste.

---

*Sooner or later, from one source or another, the student is sure to hear it said that matters of taste have no place in engineering; that time spent, for example, in making well-formed letters, or an effective title, or in attempting to make a design agreeable to the eye, as well as structurally correct, is time wasted. From a strictly commercial and utilitarian standpoint this may be true; but, on the other hand, some engineers recognize the fact that beauty in engineering works, even in machinery, is possible, and that, with a growing public appreciation, the element of beauty will not only be demanded, but paid for. Already in the building of several important bridges the question of beauty has been recognized by the engineer and architect working in conjunction. Apart from practical considerations, however, a student cannot afford, as a matter of education, to remain blind to the difference between good and bad taste.

As the question of selection is inseparable from the requirements of each particular drawing, no definite rules can be given other than the one general rule that the selection should be consistent with the purpose of a drawing. In the subject of arrangement, the following methods are important.

(a) *In designing an arrangement*, quicker and better results will be obtained if, instead of clinging to mechanical methods, the broader methods of freehand drawing be adopted. Use a rather soft pencil, and start with the intention of spoiling several sheets of paper, if necessary, to secure a satisfactory result— a good result is always worth more than the paper. When sketching in the views, let the pencil swing freely over the surface of the paper, in all directions, as suggested by the arrows, Fig. 78. Let the motion be from the arm rather than from the wrist or fingers;

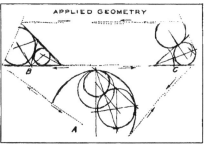

Fig. 78.

the freedom of motion thus obtained tends, in itself, to prevent one from seeing things in a detailed, constrained way. Do not begin by placing a problem up in one corner of the sheet, then another beside it, then a third one close to the second, and so on until the sheet is filled; but strike out boldly, and with the object of indicating roughly the shapes and quantity of the more important figures, in order to gain quickly a first impression of the (tentative) arrangement as a *whole*.

In working for an arrangement, it is necessary to take into account not only the shape and quantity of the problems, but also the shape and quantity of the spaces between problems, and between problems and the border line of the sheet. First place the dominant or most important problem, as *A*, Fig. 78, thereby breaking up the regular (rectangular) figure of the sheet into an irregular shape. Next place an important problem, as *B* and *C*, in a manner best calculated to fill the irregular space satisfactorily, thereby breaking up the surface into new irregular spaces, the shape and size of each of which must be considered in placing additional problems, as, for example, those shown in Fig. 79.

When a rough basis for what appears to be a satisfactory arrangement is secured, as in Fig. 79, then work out the problems with more care freehand, or with ruler and compass, but only to the extent necessary to serve as a basis for the finished sheet. In rearrangements it is convenient to make changes on tracing paper placed over the first layout, and for succeeding changes to use a fresh sheet of trac-

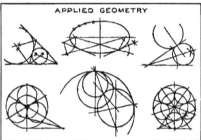

Fig. 79.

ing paper placed over the preceding one. In some cases it is convenient to cut up the sheet so that the problems — or views of an object — may be moved about until the desired arrangement is obtained, after which the loose pieces may be pasted in position.

(*b*) *Forms of arrangement.* A layout should be begun with some definite scheme or form of arrangement in mind. If it is found that the subject matter does not lend itself to a proposed scheme, other forms may be experimented with until a form is obtained which proves satisfactory.

Symmetrical lateral arrangements ; balance. If the problems — or views — on a sheet be regarded as so many actual *weights*, then, in viewing the sheet, it may be considered whether the weights on opposite sides of the center lines of the sheet counterbalance. In an arrangement symmetrical with respect to the vertical center line of a sheet, an appearance of stability may be obtained by making the problems placed on the center line larger than those placed on either side.

Symmetry with respect to a central drawing. The central drawing should be the largest on the sheet, and the other drawings should balance, both laterally and vertically, with respect to the central drawing.

Abstract arrangement. In this form the problems may be placed irregularly, but exaggerated and grotesque effects should be avoided. An effect of stability may be secured by making the weights of the problems placed low on the sheet somewhat heavier than of those placed high on the sheet — or by giving the arrangement, as a whole, the form of a truncated pyramid with slightly sloping sides.

The problems should fill the sheet ; that is, an arrangement should appear neither crowded nor scant.

In Figs. 80, 81, 82, 83, and 84 are given — as examples and for criticism — specimens of students' work in selection and arrangement, reproduced from drawings 18″ x 26″. *Criticisms : —*

Fig. 80. The sheet is sparsely filled. The left-hand two problems of the lower line are too small ; the title is too large compared with the size of the problems.

Fig. 81. The sheet is crowded.

Avoid

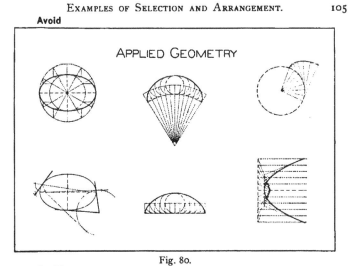

Fig. 80.

Avoid

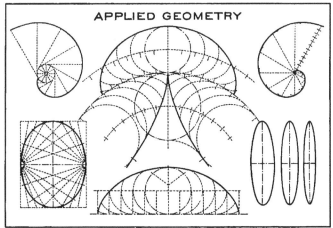

Fig. 81

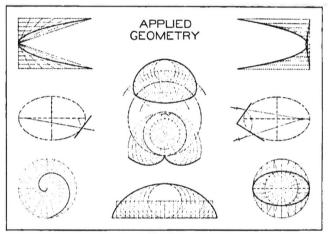

Fig. 82.

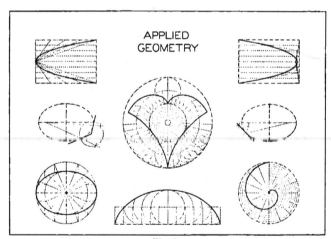

Fig. 83.

Fig. 82.   The arrangement is slightly top-heavy, the two problems at the top of the sheet being too large.   If these problems were smaller, the remaining six problems would appear too low on the sheet.

Fig. 83.   The side problems of the middle horizontal line are rather small for the central problem ; the two problems on the top line are somewhat large ; and the spacing of the side lines of problems is a trifle close compared with the spacing of the problems in the horizontal lines.

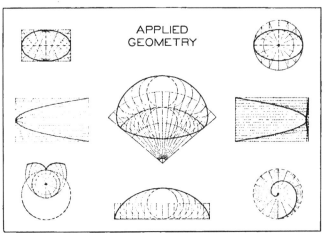

Fig. 84.

Fig. 84.   The lower left-hand problem appears farther from the bottom border line than does the lower right-hand problem.   This is due to not allowing for the emptiness of the lower portion of the larger circle ; it is the group of construction lines which here attracts the eye.

STUDY PLATE 6.

For selection, arrangement, and precise rendering.

Use Whatman's hot-pressed paper ; the sheet is to be 10″ x 14″, with a ruled border line 8″ x 12″.   It is required to make a finished plate, which shall contain problems selected from Problems 1 to 18, Chapter V., and also include one of the following titles, designed by the student : " Geometrical Construction " ;

"Selection and Arrangement"; "Applied Geometry." Make the layout on any spare paper, and do not begin the finished drawing until the arrangement has been fully decided.

PENCILING. The final sheet should be drawn very accurately; make the lines full.

INKING. Ink according to either *a* or *b*, Art. 47, and make the width of line the same as that of the lines *A*, *B*, and *C*, Fig. 55. Letter the title; also "Plate 6," your name, and the date. The letters must be drawn (Art. 41).

STUDY PLATE 7.

For selection, arrangement, and precise rendering.

Use Whatman's hot-pressed paper; the sheet is to be 14" x 20", with a ruled border line 12" x 18". It is required to make a finished plate, which shall contain problems selected from Problems 19 to 40, Chapter V., and a title. Proceed according to the directions for Study Plate 6.

STUDY PLATE 8.

For selection, arrangement, and precise rendering.

Use Whatman's hot-pressed paper; the sheet is to be 20" x 26", with a ruled border line 18" x 24". It is required to make a finished plate, which shall contain problems selected from Problems 41 to 74, Chapter V., and a title. Proceed according to the directions for Study Plate 6.

CHAPTER VII.

OBJECT DRAWING.

49. In engineering, architectural, and shop construction, objects are commonly represented by means of geometrical views called *projections*. For the complete representation of any object two or more projections are necessary, each being a different view of the object. In order to comprehend or *read* a projection drawing, it is necessary to combine the several views in imagination, thereby forming a mental picture of the object as a whole. The ability to read projections easily, usually requires extended practice, which is best obtained through the study of descriptive geometry and a considerable amount of drawing from objects.

50. Sketching and Measuring. (*a*) *The sketches.* As, in practice, objects from which drawings are to be made are usually not near by, it is necessary to make the final drawings from sketches showing the character and measurements of the object. The same procedure should be adopted when drawing from class-room models; that is, the model should be sketched and measured, and then put aside, so that the subsequent work shall be wholly from the sketches.

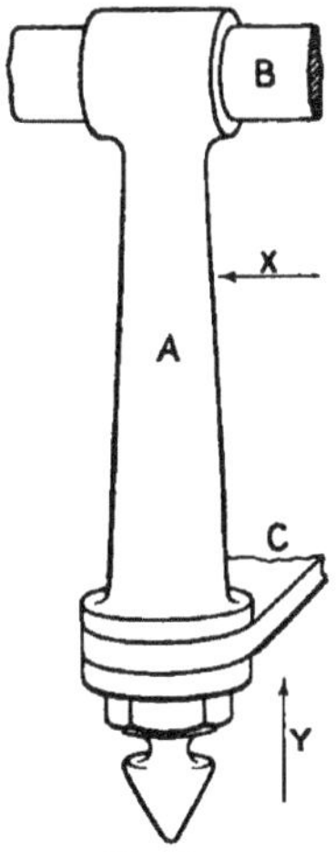

(*b*) *Character of the sketch rendering.* The sketches must not be loosely rendered, as when experimenting for an arrangement (*a*, Art. 48), but must be rendered in single line, carefully and directly — thus giving the sketch a somewhat set appearance — and with little or no erasure, or preliminary suggestion. The lines should be firm, crisp, and accurately placed with respect to the vertical and the horizontal.

(*c*) *The measurements.* Take the measurements with the two-foot rule and the calipers (see materials, p. 2). Care should be taken to make well-formed and legible numerals, signs, and arrow heads. The dimensions, other data, and remarks may be stroke rendered or neatly written. Center, extension, and dimension lines may be ruled.

Fig. 85.

All sketches made in connection with the exercises in object drawing should be placed in the note book (see materials, p. 2), and this book should be used for no other purpose.

(*d*) As a first example of sketching and measuring, take the sketch, Fig. 86, made from a wooden model of a locomotive hand-rail stud. The stud, *A* (Fig. 85),

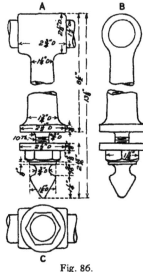

through which passes the hand rail *B*, rests on the horizontal portion, *C*, of a bracket riveted to the boiler. The necessary sketches of this object include : a front view or *front elevation* (*A*, Fig. 86) ; a side view or *side elevation* (*B*), as seen in the direction of the arrow *X*, Fig. 85 ; and a lower end or bottom view (*inverted plan*), *C*, as seen in the direction of the arrow *Y*, Fig. 85. The necessary measurements of the hand-rail stud are given in Fig. 86.

(*e*) Sketches of another drawing room model, Fig. 87, are given in Fig. 88. The object is a cabinet maker's clamp (see also Plates 13 and 14). The parts of the clamp, *A* and *B* (Fig. 87), which compress the glued pieces, *C, C, C, C*, are attached to a wooden bar, *F*. The block, *A*, is attached to the bar by the binder, *G*. The head of the clamp, *B*, is attached to the bar by the iron strap, *H*, which bears against the plate, *J*. In order that the sketches might be of a size adapted to the page of the sketch book, parts of the object were broken and brought together as shown (Fig.

Fig. 86.

88). The order in which the sketches were made is as follows : *B* (projection in the direction *Z*, Fig. 87), *B'* (projection in the direction *Y*, Fig. 87), *B''* (projection in the direction *X*) ; *A, A', D, D'*, etc. It should be noted that the same letter is used in Figs. 87 and 88 to indicate corresponding parts of the clamp.

31. **Selection and Arrangement ; the Layout.** To illustrate further methods in object drawing, let it be supposed that it is required to make a finished tracing on cloth of a mechanical drawing to be made from the sketches of the hand-rail stud, Fig. 86, and also that the tracing shall be a satisfactory example of symmetrical arrangement.* The size of the tracing is to be 14" x 20", with a ruled border line 13" x 19" (½" margin). The first step is to decide upon an arrangement.

---

* A symmetrical arrangement is chosen merely as an example ; more often an abstract arrangement (*b*, Art. 48) appears to be necessary (see Plates 13, 14, and 15).

For this purpose any spare paper is suitable; a border line of the required size should be ruled, a soft pencil used, and the procedure should be in accordance with *a*, Art. 48.

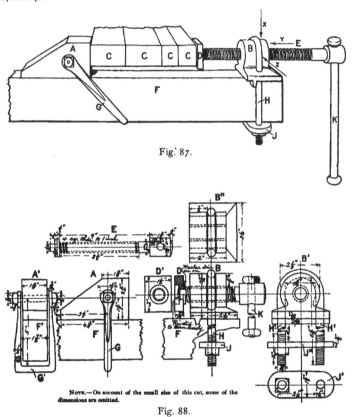

Fig. 87.

Fig. 88.

NOTE.—On account of the small size of this cut, some of the dimensions are omitted.

A glance at the overall measurements, Fig. 86, will show that, if the projections of the stud are not to appear lost on the sheet, the drawing cannot be made to a reduced scale, but must be full size. It will also be seen that the number of views, Fig. 86, necessary to represent the stud are not sufficient to fill the sheet. But in view of the requirement that, besides representing the stud, the drawing shall be a satisfactory example of arrangement, the number of projections need not be restricted, as in the case of a working drawing (*b*, Art. 64), to those projections which are actually necessary; hence a section may be added and a title introduced. A first arrangement is illustrated in Fig. 89. In this sketch the views fill the sheet

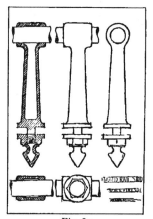 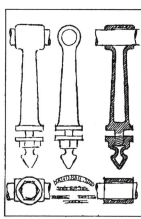

Fig. 89.   Fig. 90.

satisfactorily, but the elevation at the right-hand side of the sheet does not balance the section at the left-hand side; the title hardly balances the lower left-hand section, and the lower views are crowded. A second scheme of arrangement is shown in Fig. 90. Here the projections on either side of the central figure balance each other, but the eye is attracted to the right-hand side of the sheet on account of the heavier effect due to the cross hatching. The title and the views on either side of the title are crowded; if the size of the title is reduced to give a satisfactory spacing, it will appear too small. The third scheme, Fig. 91, the one finally adopted (see Plate 12), appears to be the best of the examples given. The stronger black-and-white value, due to the cross hatching, is kept central; and the space in the lower right-hand corner is filled by the additional section.

52. The Drawing for Tracing; Collective Rendering. (Methods continued.) The locations of the views in the pencil drawing for the tracing are taken from the layout. Of special importance is the treatment of the projections, which should be drawn in combination; that is, in penciling, no one projection should ever be carried to completion, then another completed, then another, and so on, but *at each stage of the drawing each projection should show that it has received equal attention.*

The projections of an object should always be laid out with reference to their center lines. (See *b*, Study Plate 9.)

53. Copying as a Preliminary to Object Drawing. A great deal can be learned by copying good examples of drawing, from the originals, or from prints of the same. Mere automatic imitation, however, of a drawing or print, unaccompanied by thought of the meaning and application of the construction and technique there represented, is of little value; but if the copying serves to impress correct methods of rendering, to cultivate taste and good judgment, and to afford practice in reading drawings, then there is no question that a reasonable amount of time given to copying is well spent. Students appear to be particularly weak in making neat and rapid sketches, in rendering dimensions on sketches, and in sketching layouts.

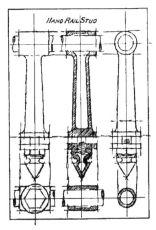

Fig. 91.

When this is the case, the student should not expect to correct his deficiency through the sketching connected with a required amount of object drawing, as the time here spent in sketching is relatively small compared with the time necessary for making the mechanical drawings. He should give special attention to sketching, and to sketch dimensioning as such, making it a practice to sketch any object at hand at every opportunity, though it be for only five minutes at a time. In this connection the preliminary exercises in sketching indicated in Study Plates 9–13 are important.

54. Screw Threads. (*a*) *Single and double threads; pitch.* If a point were carried along an edge, as *EFGHJ*, of the actual thread represented in *A*, Fig. 92, it would be seen that, in passing from point *E* to point *J*, vertically over *E*, the thread makes one turn about the axis of the screw, and that the screw has but one thread. The rise or advance, *EJ*, of the thread in making one turn is termed its *pitch*, and, in a single threaded screw, the pitch is equal to the combined widths of the thread and the groove (see *EJ*, *A*, Fig. 92).

If, in following a thread, as at *EFGHJ*, *B*, Fig. 92, an intervening thread is seen between points E and J, the screw evidently has two threads, and is said to be *double threaded.* In this case the pitch, EJ, is equal to the combined widths of the two threads and the two grooves.

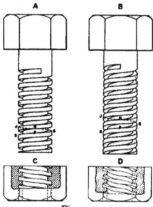

Fig. 92.

If the front of a thread ascends from left to right, as from E to J in Fig. 92, the screw is said to be right-handed; if the front of the thread ascends from right to left, the screw is left-handed.

(*b*) *The shape and size of a screw thread.* The actual shape and size of a thread are determined by the shape and size of its cross section (Fig. 93) taken in a plane passing through the axis of the screw. *The size of a thread,* while determined by the size of its section, is commonly expressed by stating the number of sections in one inch, measured parallel to the axis of the screw, thus: 10 threads to an inch (usually given in the form *10 Th.* or *10 Thds.*).

(*c*) *Measuring a screw.* When familiar with the various thread sections, Fig. 93, the shape of a particular thread can be easily identified from inspection. There should be recorded in the sketch: the shape of the thread; the number of threads to an inch; and whether the thread is single or double. The necessary measurements of a threaded bolt are as follows: total length, including the head; length of body, including threaded portion; length of the threaded portion; diameter of body; height and short diameter of the head.

(*d*) *Drawing a screw thread.* The drawing must be begun by making a longitudinal section A, Fig. 94, taken through the axis of the screw. For a complete representation, it is necessary to show the curvature of the thread (see helix, descriptive geometry); but ordinarily, in object drawing, straight lines, as *FH, HM, JK,* and *KL, B,* Fig. 94, are substituted for the curves.

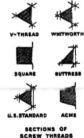

SECTIONS OF
SCREW THREADS

Fig. 93.

In each of Figs. 94–97 are illustrated four things, namely: at A, the pencil construction; B, the directions of the front and back edges of the thread; C, the front of the completed thread; and D, the completed thread as seen in a section of a tapped hole or nut.

(*e*) *Construction of a single, right-hand, V thread, 4 threads to an inch.* (Fig. 94.) Lay off *EG* equal to the given diameter of the screw, and draw the contour elements, *EF* and *GH* — the entire length of the threaded portion. Set the scale to the element *EF*, and, starting at point *E*, lay off $\frac{1}{4}''$ distances, as 1–2, 2–3, the entire length of the threaded portion. With the 30° triangle, draw the section, 1*R*2, of the groove, and through point *R* draw the line *RS* parallel to *EF*. Complete the sections of the groove and thread along the left-hand side of the screw, and, as a check, see that the roots, as *R* and *S*, of the thread fall accurately in the line *RS*. From point *R*, draw *RP* perpendicular to line *EF*, intersecting element

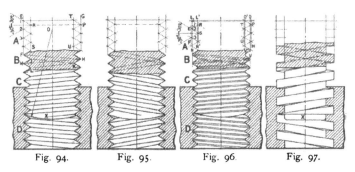

Fig. 94.   Fig. 95.   Fig. 96.   Fig. 97.

GH in point *P*, thus locating — since the thread is single — a *point* of the thread. Starting at point *P*, draw one triangle representing a section of the thread, and also the line *TU* containing the roots of the thread. Lay off from the scale the positions of the points of the threads, and complete the sections on the right-hand side of the screw. Having thus drawn the sections of the thread, note (at *B*), the one turn *FHM*, and *JKL*, of the point and root edges respectively; then draw the visible edges of the screw, and of the nut, as shown at *C* and *D*. The end of the screw in the drawing is usually finished by an arc of a circle, as *X*, Fig. 94, described from any appropriate point, as *O*, lying in the axis of the screw.

(*f*) *Construction of a double, left-hand, V thread, 2 threads to an inch.* (Fig. 95.) Observe that the pitch is $\frac{1}{2}$ inch, and that, since the thread is double, the points of the sections on opposite sides of the screw fall opposite each other; otherwise the construction is the same as in Fig. 94.

(*g*) *Construction of a U.S. standard thread, 4 threads to an inch.* (Fig. 96.) Draw the contour elements, *LN* and *OV*, of the screw. Outside of *LN*, and *OV*, at a distance equal to $\frac{1}{8}$ of the pitch of the screw, draw *EF* and *PH*. Using

EF and *PH* as corresponding to *EF* and *PH* in Fig. 94, construct the sections of a V thread of the given pitch (see *c*). Outside of *RS*, and *TU*, at a distance equal to ⅛ the pitch of the screw, draw *L'N'* and *O'V'*. Connect the points of the sections as shown at *C* and *D*, Fig. 96.

(*h*) *Construction of a single, square thread, 4 threads to an inch.* The construction is evident from Fig. 97.

55. Conventional Screw Threads. Any of the conventions shown in Figs. 98 and 99 may be used according to preference or requirement. The constructions are as follows : —

(*a*) *Convention A, Fig. 98.* Starting at point *D*, lay off, for the entire length of the left-hand contour element, any assumed distance, as *DF;* at *E*, opposite *D*,

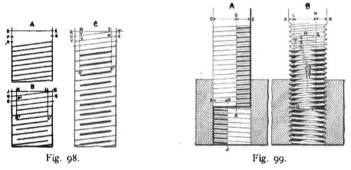

Fig. 98.   Fig. 99.

lay off the same distance, *EG*, on the right-hand contour element. Connect the points *F* and *E ;* then, by sliding the triangle, draw lines parallel to *FE*, passing through the points laid off on the left-hand element.

The distance laid off on the elements need not be the actual pitch of the thread, but any convenient distance which gives a satisfactory spacing for the parallel lines.

(*b*) *Convention B, Fig. 98.* First draw the thread in pencil according to *A*, Fig. 98. Make the distances *JK* and *LM* each equal to the unit, as *JN*, of the vertical spacing, and draw *KK'* and *LL'*. When inking the convention, begin and end the lines representing the thread, alternately, at the lines *KK'* and *LL'*. Thus, for example, draw from point *N* to line *LL'*, and from point *P'*, in *KK'*, to line *MQ* (see the lower portion of the figure).

(*c*) *Convention C, Fig. 98.* First draw the thread according to the pencil construction in *B*, Fig. 98. Ink the convention as shown in the figure. The wide lines should be first represented by two narrow lines, and afterward filled in as described for the wider lines in section lining (*b*, Art. 38).

(*d*) *Convention A, Fig. 99.* The pencil construction is the same as for convention *B*, Fig. 98. Make *DE* and *FG* each equal to $\frac{3}{8}$ of *CE*, and draw lines *DH* and *GJ*. Ink the convention as shown in the figure.

(*e*) *Convention B, Fig. 99.* The thread should be drawn according to *e*, Art. 54. The black areas represent conventional shadows, the curved edges of which may be located as follows : Lay off distance *KL* equal to $\frac{1}{4}$ of *KN*, and draw line *LU*. Make *MN* equal to $\frac{1}{3}$ of *LN*, and draw *MV*. Through the points, as *P* and *Q*, in which *LU* and *MV* intersect the root of a thread, draw a circular arc, center found by trial, tangent to the point of the thread, as at *O*. Through the center, *R*, of this arc, draw line *OT*. With the same radius, and keeping the center always in line *OT*, draw the arcs for all the shadows, as indicated by the arrows. Fill in the shadows with India ink applied with a brush or a writing pen.

56. Chamfer of Nuts and Bolt Heads. Let Fig. 100 represent the head of a bolt supposed to revolve about its axis *ab*, while the tool, *Z*, cuts off or *chamfers* the edges, *JQ*, *QX*, *XR*, etc. The surface *BFNUWTP*, etc., of the chamfer is a portion of the surface of the imaginary cone, *def*: the right-hand edge, *FNU*, of the chamfer is a circle, since it is the boundary of a right section of the cone *def*; and the left-hand boundary *BEHMW*, etc., is a series of six equal hyperbolic arcs, being the intersection of the cone *def*, by the six faces of the bolt head. Since the base of the head is perpendicular to the axis, *ab*, the edges *AB*, *GH*, *OP*, and *VW* are cut off the same length; likewise the distances *DE*, *LM*, and *ST* are equal. In practical drawing, the hyperbolic curves are always represented by arcs of circles; the position of the arcs may be determined as follows : —

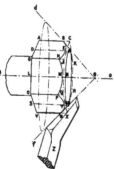

Fig. 100.

(*a*) *Construction.* Through the lowest point, *A*, Fig. 101, in the chamfer (located by measuring the edge *AB*), draw the line *AD* making, in this case, 45° with the top surface of the bolt head. With radius *CD*, draw the circle *D'G'*, representing the top edge of the chamfer. The lowest points in the chamfer, as *E*, *A*, and *E''*, lie in the vertical edges of the head. The highest points in the chamfer, as *H*, *N*, and *N''*, lie in the center lines of the vertical faces. Pass the plane *C'K'J'*, containing the axis of the head and bisecting its front face; this plane, revolved, corresponds to the half side view *C''K''J''*. Lay off *C''G''* equal to *C'G'*, and

through point G'' draw the line $G''H''$, at 45°, thus determining point H'', which is
the highest point of the chamfer, lying in the center line HJ of the front face of

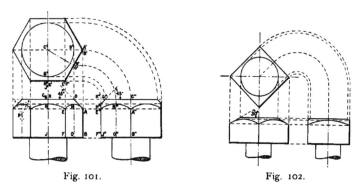

Fig. 101.   Fig. 102.

the head. Project the highest and lowest points in the curves as shown. Find
each center, as P, from which to describe the circular arcs, according to b, Art. 45.

A 30° conical chamfer on a square head is shown in Fig. 102. A spherical
chamfer is shown in Fig. 103. It will be seen that the chamfer is made only deep

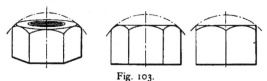

Fig. 103.

enough to complete the curve on each face of the nut; the lower edges of the
chamfer are therefore tangent to its upper edge.

57. Doubtful Lines. In object drawing it frequently happens that, on account
of rounded corners or other curved surfaces, a part of an object shows no definite
line boundary; hence, theoretically, the part cannot be expressed by an outline.
When, in such cases, the question arises whether a line should be put in or omitted,
it is best to ignore geometrical truth, if the drawing will be made clearer thereby.
Examples : —

(a) *Detail of the clamp, A, B, and C, Fig. 104. (See also Plate 13.)* The
strictly correct projection, B, can show no interior lines, since all corners of the

object are slightly rounded, as shown in the boundaries of *A* and *B*.   The more satisfactory drawing, *C*, is obtained by putting in the doubtful lines.

(*b*)   *The angle iron, D, E, F, Fig. 104.*   The projection *D* is correct ; but the addition of the doubtful lines in the projection *E* makes this view the clearer.

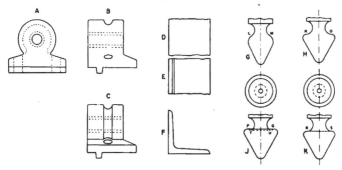

Fig. 104.

(*c*)   *The ornamental terminations of a hand-rail stud, G, H, J, K, Fig. 104.* (*See also Plate 12.*)   In the form *G*, it is clear that there should be no line drawn from *L* to *M* ; in the form *J*, it is equally clear that there should be a line from *P* to *Q*.   Forms *H* and *K* are the same ; the former, without a line drawn from *N* to *O*, is correct ; but the latter, showing the doubtful line, *RS*, seems preferable.

58.   Shade Lines.   If rays of light from any source be supposed to fall on an object, some of its faces will be in the light, the rest in shadow.   The edges of the object that separate its light from its dark surfaces are called shade lines.   These lines are indicated in a drawing by their width, which is greater than that of any other lines in the same drawing.   Shade lines may be used to make a drawing more effective in appearance, and, to a limited extent, to explain form.

(*a*)   *Theory of shade lines ; rectangular objects.*   Let *B*, Fig. 105, represent a cube with its base horizontal.   The cube is supposed to be lighted by parallel rays, the direction of which is represented by the diagonal, *ed*, of the cube (arrow *X*), which makes the angle, *β*, of 35° 15′ 52″ with all faces of the cube.   It will be seen that the faces *abfe*, *efhg*, and *aegc* receive the light, while the remaining faces are in shadow ; hence the shade edges of the cube are those forming the boundary *abfhgc*.   Next, let the cube be represented by the projections *A* and *C* (elevation and plan).   To determine the shade lines of the projections, it is necessary also to represent the light by means of projections of its rays.   The elevation X^v (B) of the ray X, coincides with the diagonal, *ad*, of the back face of the cube, and therefore takes a direction downward and to the right at 45° with the horizontal, and is so represented in the elevation, A, of the cube. The plan, X^A (B) of the ray X, coincides with the diagonal, *gd*, of the base of the cube, and therefore takes a direction backward and to the right at 45° with the horizontal, and is so represented in the plan, C, of the cube. Now, by reading in combination the elevation and plan of both object and rays, the shade lines may be determined. Thus, for example, the elevation (A) shows that the top surface, *abfe*, of the plan (C) is in light; the plan shows that the back surface, *abdc*, is in shadow; hence the line *ab*, representing the (shade) edge separating the two surfaces in question, is a shade line.

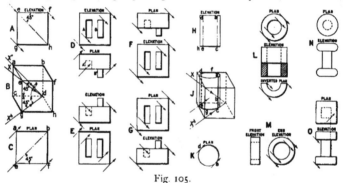

Fig. 105.

A cylinder. Let *J*, Fig. 105, represent a cylinder, with its base horizontal, inscribed in a half cube, as indicated by the diagonals X and X^A. The upper base, *ad*, is in the light; the lower base is in shadow. The ray of light, X, is tangent to the curved lateral surface at point *b*; therefore, the element *ac*, passing through point *b*, separates the light from the dark portion of the curved surface, and is a shade line. The element *de*, diametrically opposite *ac*, must also be a shade line; hence, the entire shade-line boundary of the cylinder is the broken line *afdegca*.

As a preliminary to object drawing, further consideration of the theory of shade lines is unnecessary. To determine all shade lines, in all cases, requires an extensive knowledge of the theory of shades and shadows — a subject useful in certain architectural drawing and for the training it affords, but of no practical use in construction drawing, since the process requires much time, and in many cases the shade lines become so complicated that they are more likely to obscure than to explain the form of an object.

In practice there is no general understanding in regard to shade lines; the architect shades his drawings in one way, the engineer in another. Then, draftsmen differ in opinion as to whether particular lines should be shaded or not, and this is an ever-present theme for discussion. In any case, the question to be decided is at what point the theory shall be ignored in the interest of clearness and utility. In some offices, shade lines are omitted altogether. The following examples represent the more general practice : —

(b) *Architect's method of shade lines.* Figs. *D, E, H,* and *K,* Fig. 105, are shaded according to the architect's method. The rectangular objects, Figs. *D* and *E,* are shaded strictly in accordance with the shading of the cube (*A* and *C*). In *D,* the shade lines on the elevation show that the rectangle, *A,* represents a recess, and the rectangle, *B,* a projecting block. In *E,* the shade lines on the plan indicate the same facts. The elevation of the cylinder (*H*), is not shaded wholly in accordance with theory : — The shade elements, *ac* and *dc,* do not coincide with the contour (outside) elements of the cylinder ; they are therefore omitted, and no element is shaded. The shade line of the lower base, by theory extending only from *h* to *c,* is carried clear across the base, and no part of the upper base is shaded. The shade line of the plan (*K*), agrees with the theory.

(c) *Engineer's method of shade lines.* In this method, the direction of the light is taken the same for all views and sections of an object. The direction assumed is downward to the right, and all views are shaded similarly to the elevation in the architect's method. The shading is generally done by some arbitrary rule, such as the following : Shade the lower and right-hand (sharp) edges of objects, and the upper and left-hand (sharp) edges of holes on all views and sections. A "sharp" edge is usually taken to mean, besides an actual angle, a rounded corner where the rounding is very slight, and intended only for a finish. Examples : —

In Figs. *F* and *G,* Fig. 105, the same objects as in *D* and *E* are shaded according to the engineer's method. A comparison will show the difference between the two methods of shading.

An elevation, a section, and two plans of a hollow cylinder are given in *L.* The right-hand contour element of the full cylinder is not shaded because it is not a "sharp" edge ; but, where the section is taken, both the right-hand edge of the cylinder and the left-hand edge of the hollow are shaded.

The object, *N,* as it is bounded entirely by curved surfaces, can have no shade lines, since none of its edges is "sharp."

The object, *O,* is bounded by flat surfaces ; but the corners are slightly rounded for a finish. In this and similar cases, by theory, the shade lines would be represented by the fine dash-and-dot lines shown in the figure, and hence the contour would not be shaded ; in practice, however, in order to indicate the general (rectangular) form of the object, the edges might be shaded.

(*d*) The following rules apply to both systems.

Details which fall within the shadow of an object of which they are parts should be shaded with respect to their own light and shade, unaffected by that of the larger object. Thus, for example, each link of the bicycle chain, Fig. 106, is

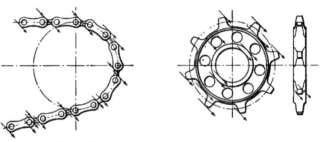

Fig. 106.   Fig. 107.

shaded according to the arrows shown, whether or not the sprocket wheel is in place. The rim of the sprocket, Fig. 107, casts a shadow on some of the teeth; but all of the teeth are shaded individually according to the arrows.

Dotted lines, representing invisible edges of an object, *are never shaded*.

Finally, if there is any question as to whether a line should be shaded or not it is usually best to leave it unshaded.

59. Representation of Irregular Objects; Mixed Rendering. An object is said to be irregular when it cannot be readily resolved into the common geometrical forms, such as prisms, cylinders, spheres, etc. To represent an irregular object, it is necessary to locate geometrically its essential points, which are then connected by lines rendered, according to convenience, with the instruments, by means of the French curve, freehand, or by a combination of these methods (mixed rendering).

(*a*) *Methods of measuring irregular objects.* Either of the following methods may be used in locating points in an object. (I.) *By base lines and offsets (rectangular co-ordinates).* Assume or locate a base line, such as the line AB of the hook, Fig. G, Plate 11. From the point required to be located, as K, let fall a perpendicular or offset, as KH, to the base line. Locate the position of the foot of the offset by measuring along the base line, and then measure the length of the offset. (II.) *By triangulation.* To locate a point, as D, Fig. E, Plate 11, measure the distances, as AD and BD, from each end of any line, as AB, already determined.

Plate II
(Study Plate 14)

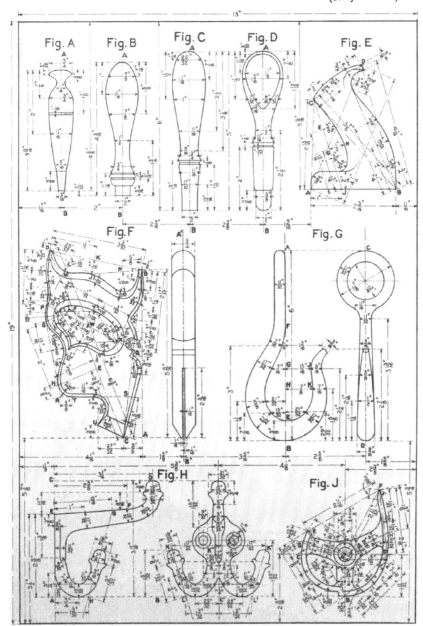

The dimensions and reference letters should not appear on the student's drawing.

If an object is quite irregular, the measuring may require considerable judgment and ingenuity.

(*b*) *A three-pronged hook.* To illustrate modes of procedure, take the three-pronged hook, Fig. *H*, Plate 11, also shown pictorially in Fig. 108. The hook is symmetrical with respect to an imaginary plane (*central plane*) which passes through the highest point, *D*, of the hook, and contains the center line of the upper prong, *EGD*. The imaginary plane *RSZ'Y'* is parallel to the central plane; and the imaginary plane *QRY'X'*, representing the flat surface of the hook, *Z*, produced, is perpendicular to plane *RSZ'Y'*.

The left-hand view, Fig. H, Plate 11. This is a view seen in the direction of the arrow *X*, Fig. 108, perpendicular to the imaginary plane *RSZ'Y'*; hence all measurements for this view must be taken in directions parallel to the plane *RSZ'Y'*. In Fig. 108 the central plane of the hook is represented by the triangles *1'*, *2'*, and *3*. The total height of the hook is equal to the distance *CA'*, measured along the edge of the triangle *1'*, between the edge *CD* of the triangle *3*, which passes through the highest point, *D*, of the hook, and the edge *A'B'* of the triangle *2'*, which passes through a point midway between the lowest points *A'''* and *A^{iv}*

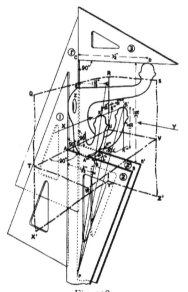

Fig. 108.

of the hook. Point *A'''* is located by the base line obtained with the triangle *1*, and the offset *AA'''* established with the triangle *2*.

The upper prong, since it is parallel to plane *RSZ'Y'*, can be wholly located by means of points, such as *D*, determined by base lines and offsets. For the lower prongs, *K*, *K'*, a different method must be adopted, as the center line of each prong lies in a plane *oblique* to both *RSZ'Y'* and *QRY'X'*. The upper ends of the prongs, being comparatively straight, may be referred to the center lines *KL* and *K'L'*. Viewed in the direction of the arrow *X*, the nearer prong hides the further one, and, therefore, only the nearer one need be considered. In order to locate two points in the center line, GP, a means must be found for projecting the points on lines which are parallel to the plane $RSZ'Y'$. The following method is usually sufficiently accurate. Looking in the direction of the arrow X, hold any straight-edge parallel to plane $RSZ'Y'$, between the eye and the hook, so that the edge will cover the center line. Without changing the position of the eye or of the straight-edge, note carefully, either by sighting or by squaring out with the triangle held perpendicular to the edge, the points where the straight-edge *appears* to cross any two lines which are parallel to the plane $RSZ'Y'$. For example, points so located are point G in the upper prong, and point H in the offset AB.

The curved portion of the prongs may be determined by measuring diameters, as $\frac{1}{32}''$ and $\frac{7}{16}''$, Fig. H, Plate 11, and then locating the position of these diameters by base lines and offsets as shown.

The right-hand view, Fig. H, Plate 11. This is a view seen in the direction of the arrow Y, Fig. 108, perpendicular to the imaginary plane $QRY'X'$; hence all measurements must be taken in lines parallel to this plane. *No dimensions, however, should be taken which can be projected from the left-hand view, Fig. H.* For example, the measurement $1\frac{5}{32}''$, Fig. 108, giving the vertical height of the lower prong, is wholly unnecessary, as this height is already determined by means of the center line JH, in the left-hand view of Fig. H. The projection of the angle, KPK', Fig. 108, made by the center lines of the straight portions of the lower prongs, is readily determined by measuring the lines KK' and LL' parallel to the plane, $QRY'X'$, the heights of which are projected from the left-hand view, Fig. H.

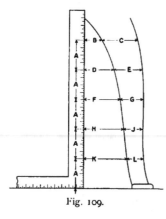

Fig. 109.

(c) *Turned handles.* (Figs. A, B, C, and D, Plate 11). Use the center line, AB, of the handle as an axis. Measure diameters at a sufficient number of points to determine the curvature fully, and locate the positions of these diameters by measurements taken parallel to the axis. Endeavor, as far as possible, to find places where the measurements will come some even division on the scale, for both the diameters and the distances along the axis. In addition to the measurements *the distinctive characteristics of the curves should be sketched as accurately as possible.*

(d) *Oblique curves.* In locating the points of a curve which takes a diagonal direction, a carpenter's square may be used to advantage. Let the curves, Fig. 109, represent the edges of the legs of a lathe or other machine, which are oblique to a vertical surface of the bed or table portion of the machine. Place one edge of the square on the floor, with the other edge passing through the highest point of the left-hand curve, Fig. 109. On the floor draw a line passing through the foot of the vertical edge of the square and perpendicular to the front vertical surface of the bed of the machine. Keeping the lower end of the vertical edge of the square always in the line on the floor, move the square forward so that each measurement, as *B* and *C*, *D* and *E*, etc., will lie in a vertical plane containing the face of the square. Obtain in a similar manner a projection in a direction at right angles to that of the first projection.

(*e*) It is sometimes convenient to duplicate a given curve by cutting out a template, or by fitting a strip of sheet lead to the curve ; the curve is plotted from the template or lead instead of measuring the original.

(*f*) In concluding this subject, it should be pointed out that the drawing of irregular objects is a final test of skill in rendering, especially when the representation depends largely on freehand methods. The common objects given in Plate 11 are typical of the more important cases which are likely to arise, and afford excellent practice in mixed rendering and in the management of complicated dimensions. Although objects requiring treatment similar to those on Plate 11 may be met with only occasionally in engineering drawing, yet, when such occur, they should be as well rendered as the purely instrumental work ; otherwise the appearance of the whole drawing will be ruined.

STUDY PLATE 9.

For sketching; the rendering of dimensions on sketches; collective rendering; reading drawings; lettering and dimensioning; speed.

(*a*) *Sketching.* Place Plate 12 from 20″ to 30″ from the eye and perpendicular to the line of sight. On a single page of the sketch book (see materials, page 2) make freehand sketches of Figs. *A, B, C, D,* and *E,* Plate 12 ; *the sketches must not be the same size as the views on the plate.* Use an H to 3H pencil, and let the sketches fill the page. Read Art. 50 ; plan the general position of the several views on the page ; lay in the masses of each projection, and then add the details. Place on the sketch all measurements and data given on Plate 12, except when marked * ; render the numerals neatly and rapidly, and without the aid of guide lines.

(*b*) *Collective rendering.* It is required to make a pencil drawing of the handrail stud, on duplex detail paper, for a tracing the size of which is to be 14″ x 20″, with a ruled border line 13″ x 19″ ($\frac{1}{2}$″ margin). Make this pencil drawing from the above sketches and measurements, but locate the views according to the measurements given on the Plate 12. Lay out the projections as follows: Draw the center lines, *NO*, *FQ*, *RS*, *TU*, and *VW*, Fig. 110. Do not complete one projection independently of the other projections, but, within practical limits, *let each stage of the drawing show that each projection has received equal attention.* Thus, for example, draw the circle *A* (*K*, Fig. 110), and project the width of the parts *A'* (Fig. *J*) and *A''* (Fig. *H*) from this circle. With the compass as set for circle *A*, lay off the widths of the part *A'''* (*L*) and *A⁴* (*M*). Take in the dividers one-half the length of part *A'*, Fig. *J*, and with this setting, measuring from the center lines, lay off the lengths of *A'*, *A''*, *A'''*, and *A⁴*. Draw the circle *B* (Fig. *L*); from *B* project *B'*, *B'*, and with the compass lay off the width of *B'* at *B''*, *B''*, and *B'''*, *B'''*. Draw the nut *E* (Fig. *L*), according to *s*, Art. 45; project *E'* (Fig. *H*); make *E''* equal to *E'*. Make the width of the nut at *F* equal to *FG* (Fig. *L*). From this point the several views may be treated more in detail.

Draw the screw threads according to *c*, Art. 55.

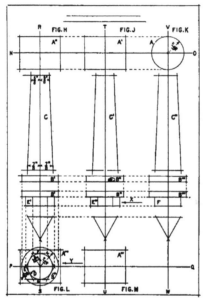

Fig. 110.

(*c*) *The tracing; the lettering and dimensioning.* Trace the drawing according to the general directions for tracing, Study Plate 1 (also see *e*, Art. 44). Make the widths of line according to those of the lines *D—J*, Fig. 55. Take particular notice of the arrangement, on Plate 12, of the dimensions and data relative to each view of the object; then stroke render all dimensions and data, not marked *, given on the plate; make the style and size of the numerals and letters according to Fig. *F*, Plate 5. Balance the title (*b*, Art. 41) on the vertical center line of the sheet; make the letters of the title according to the style and heights given on Plate 12; the letters should be *drawn* (*a*, Art. 41).

Hand in the sketch book, the pencil drawing, and the tracing.

Plate 12
(Study Plate 9)

HAND RAIL STUD

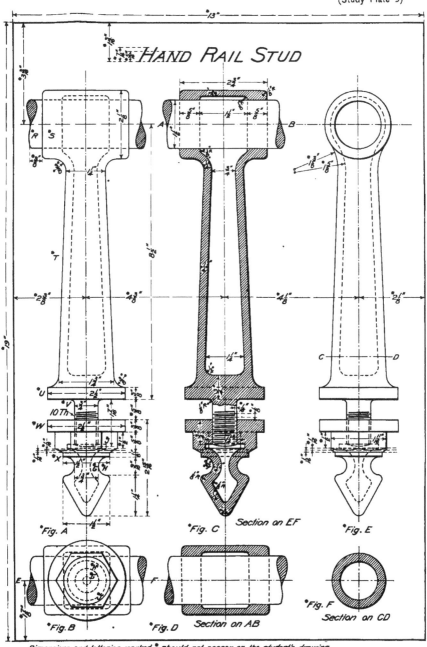

*Fig. A

*Fig. B

*Fig. C

Section on EF

Section on AB

*Fig. D

*Fig. E

*Fig. F

Section on CD

*Dimensions and lettering marked * should not appear on the student's drawing*

STUDY PLATE 10.

For sketching; the rendering of dimensions on sketches; arrangement; collective rendering; reading drawings; lettering and dimensioning; speed.

(*a*) *Sketching.* Sketch the several projections, Plate 13†; proceed strictly in accordance with *a*, Study Plate 9, but let the sketches fill *two* (or *three*) pages of the sketch book; *the sketches must be larger than the views on Plate 13.* Break the screw and bring the parts together (*b*, Art. 37). Place on the sketches all measurements and data given on the plate, except when marked *.

(*b*) *Arrangement.* On any spare paper, rule a border line 13″ x 19″, and design an arrangement different from that of Plate 13, but which shall include all of the views there given. (See Art. 51.)

(*c*) *Collective rendering.* Use duplex detail paper; rule a border line 13″ x 19″; and make, *from the sketches* (see *a*), a drawing which is to be traced. The views must be located according to the student's layout (see *b*). Draw the views with respect to their center lines, and according to the method suggested in *b*, Study Plate 9. Draw the square threaded screw according to Fig. 97.

(*d*) *The tracing; the lettering and dimensioning.* The size of the tracing is to be 14″ x 20″, with a ruled border line 13″ x 19″. Proceed according to the general directions for tracing Study Plate 1 (also see *e*, Art. 44). Make the widths of line according to those of the lines *D—J*, Fig. 55. Take particular notice of the arrangement, on Plate 13, of the dimensions and data relative to each view of the object; then stroke render all dimensions and data, not marked *, given on the plate; make the style and size of the letters and numerals according to Fig. *F*, Plate 5. Balance the title on the vertical center line of the sheet; make the letters of the title according to the styles and heights given on Plate 13; the letters should be *drawn* (*a*, Art. 41).

Hand in the sketch book, the layout, the pencil drawing, and the tracing.

STUDY PLATE 11.

For the assembling of details; speed.

Use Whatman's cold-pressed paper. The finished sheet is to be 14″ x 20″, with a ruled border line 13″ x 19″.

Make an assembly drawing of the head of the clamp, from the sketches, *a*, Study Plate 10. Locate all views, and draw the projections of the block of the clamp according to the measurements given on Plate 14.

Ink the drawing carefully, according to *e*, Art. 44; make the widths of line correspond to those of the lines *D*, *E*, *F* and *G*, Fig. 55. The dimensions and data, not marked *, should be stroke rendered; make the style and size of the letters and

† The clamp and the wrench are from non-related objects.

numerals according to Fig. *F*, Plate 5. Balance the title on the vertical center line of the sheet ; make the letters of the title according to the styles and heights given on Plate 14 ; the letters should be *drawn* (*a*, Art. 41).

STUDY PLATE 12.

For sketching; the rendering of dimensions on sketches; arrangement; collective rendering ; reading drawings; lettering and dimensioning; speed.

(*a*) *Sketching.* Sketch the details of the hanger, Plate 15 ; proceed strictly in accordance with *a*, Study Plate 9, but let the sketches fill *three* (or *four*) pages of the sketch book ; the sketches *must be larger than the views on Plate 15*. Place on the sketches all measurements and data given on the plate, except when marked *.

(*b*) *Arrangement.* On any spare paper rule a border line 13″ x 19″, and design an arrangement different from that of Plate 15, but which shall include all of the views there given. (See Art. 51.)

(*c*) *Collective rendering.* Use duplex detail paper ; rule a border line 13″ x 19″. Make, *from the sketches* (see *a*), a drawing which is to be traced. The views must be located according to the student's layout (see *b*). Draw the views with respect to their center lines, and according to the method suggested in *b*, Study Plate 9. Draw the large V-threaded screw according to *e*, Art. 54.

(*d*) *The tracing ; the lettering and dimensioning.* The size of the tracing is to be 14″ x 20″, with a ruled border line 13″ x 19″. Proceed according to the general directions for tracing Study Plate 1 (also see *e*, Art. 44). Make the widths of line according to those of the lines *D—J*, Fig. 55. Take particular notice of the arrangement, on Plate 15, of the dimensions and data relative to each view of the object ; then stroke render all dimensions and data, not marked *, given on the plate ; make the style and size of the letters and numerals according to Fig. *E*, Plate 5. Locate, vertically, the lines of the title according to the measurements given on the plate, but center the title between the edge of the drawing and the ruled border line. The letters of the title should be *drawn* (*a*, Art. 41), and their style copied from Plate 15.

Hand in the sketch book, the layout, the pencil drawing, and the tracing.

STUDY PLATE 13.

For the assembling of details ; speed.

Use Whatman's cold-pressed paper ; the finished sheet is to be 14″ x 20″, with a ruled border line 13″ x 19″.

Make an assembly drawing of the hanger from the sketches, *a*, Study Plate 12. Locate the drawings according to the measurements given on Plate 16.

Plate 13
(Study Plate 10)

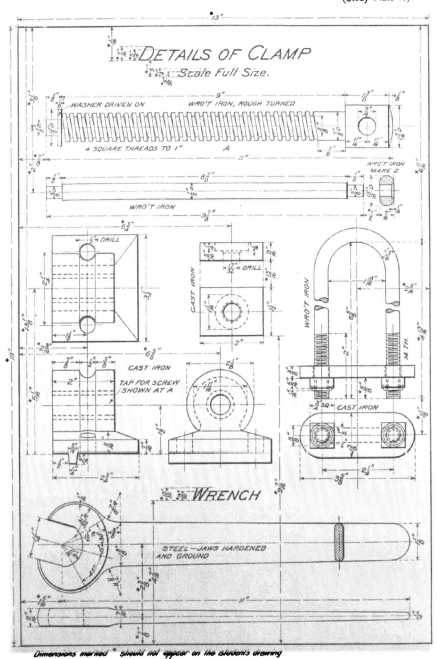

DETAILS OF CLAMP
Scale Full Size.

WASHER DRIVEN ON   WRO'T IRON, ROUGH TURNED

4 SQUARE THREADS TO 1"   A

WRO'T IRON
MAKE 2

WRO'T IRON

DRILL

CAST IRON

DRILL

CAST IRON

TAP FOR SCREW
(SHOWN AT A)

WRO'T IRON

14 TH.

SQ.   CAST IRON

WRENCH

STEEL — JAWS HARDENED
AND GROUND

Dimensions marked * should not appear on the student's drawing

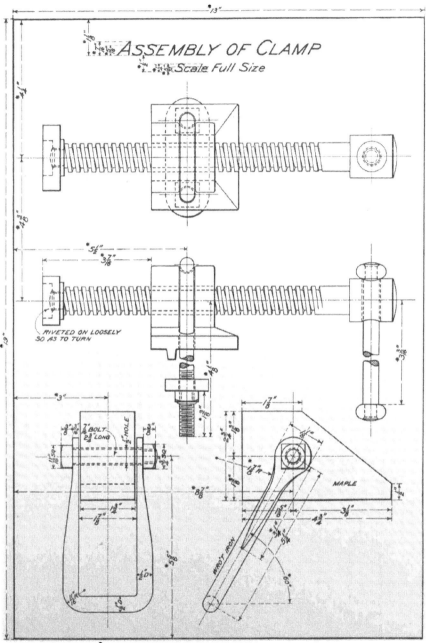

Plate 14
(Study Plate 11)

ASSEMBLY OF CLAMP
Scale Full Size

RIVETED ON LOOSELY
SO AS TO TURN

BOLT
LONG
HOLE

MAPLE

WROUGHT IRON

Dimensions marked * should not appear on the student's drawing

Plate 15

(Study Plate 12)

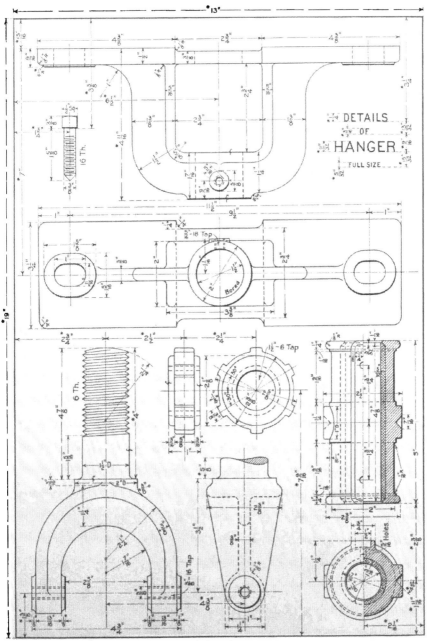

DETAILS
OF
HANGER
FULL SIZE

Dimensions marked * should not appear on the student's drawing

Plate 16

(Study Plate 13)

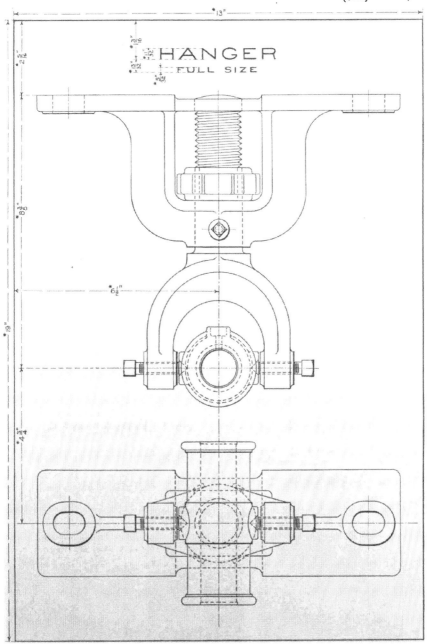

HANGER

FULL SIZE

Dimensions marked * should not appear on the student's drawing

(139)

Ink the drawing carefully according to e, Art. 44; make the widths of line correspond to those of the lines D, E, F, and G, Fig. 55. Balance the title on the vertical center line of the sheet; make the letters of the title according to the style and heights given on Plate 16; the letters should be *drawn* (a, Art. 41).

STUDY PLATE 14.

For practice in managing complicated measurements; mixed rendering in pencil and in ink.

Use Whatman's hot-pressed paper; the finished sheet is to be 14″ x 20″, with a ruled border line 13″ x 19″.

I. PENCILING. Lay out the border line, and draw the several figures full size, according to the dimensions given on Plate 11.

(a) *The turned handles (Figs. A, B, C, and D).* Locate and draw, of indefinite length, the center lines AB. Locate the lower end, and lay off the height of the handle. On the center or base line lay off the distances which locate the positions of the diameters; lay off the diameters on lines drawn perpendicular to the base line. After all the points in the contour of the handle have been located, connect them by a lightly rendered freehand line. To test the symmetry of the contour of each handle, place tracing paper over the drawing, rule the center line, and carefully trace one side of the contour. Turn the tracing paper over, make the two center lines coincide, and compare the semi-contour on the tracing paper with the underlying semi-contour of the original. The original may be corrected by means of the tracing paper, but the points located by scale measurement *must not be changed.*

(b) *The plane handle (Fig. E).* Locate AB, and, using this line as a base, triangulate for point D thus: With radius $4\frac{1}{16}''$, center B, describe an arc; with radius $4\frac{1}{4}''$, center A, intersect the preceding arc in point D. Draw BD and AD. With AB as a base, in a similar manner triangulate for point C. Locate point J, lying in AB, $1''$ from point A. Draw the base line CJ; locate the offsets EF and GH; measure the offsets, and connect the points C, F, and H. Draw a horizontal line $\frac{9}{16}''$ above AJ. With radius $\frac{9}{32}''$, center in a line drawn parallel to, and $\frac{5}{32}''$ distant from the base line CJ, draw a circular arc tangent to the preceding horizontal line above AJ. Produce, freehand, the line already drawn through points F and H, to give a smooth curve tangent to the circular arc. Proceed in a similar manner, and, in general, *consider the longer distances first.*

(c) *The saw handle (Fig. F).* Locate and draw the vertical line AB. Draw AC perpendicular to AB; locate point C, and draw CB. With CB as a base line, triangulate for point D. With DC as a base line, locate points G and H by means of the offsets FG and EH. Locate point R, lying in GH produced; locate point S in line CB, and draw RS. Draw base line DB, and locate point J by means of the offset *KJ*. Locate point *P* on line *DG*, point *Q* on line *BC*, and draw base line *PQ*. Locate points *V* and *X* on line *PQ*, point *Y* by means of the offset *WY*, and draw *VY* and *YX*. Then, proceeding as thus suggested, finish the *traverse* or generalized boundary of the handle *as a whole ;* locate, by means of base lines and offsets, the more important points in the boundary ; and finally, locate and draw the curves in detail.

(*d*) *The hooks and the cam (Figs. G, H, and J).* The drawings of these objects should be plotted according to the preceding methods.

II. INKING. Make the width of line equal to that of line *D*, Fig. 55. *All freehand lines* and lines ruled by means of the French curve should be so skillfully rendered that they will be uniform in appearance with the straight lines made with the ruling pen.

(*e*) *Lettering.* Letter "Plate 14," your name, and the date. The dimensions and all other lettering may be omitted.

CHAPTER VIII.

60. A Working Drawing is a drawing, made in accordance with engineering or architectural practice,* which presents such views and measurements of an object as will enable a mechanic to make the object wholly from the drawing.   For complicated structures, such as buildings, bridges, and machines, two kinds of drawings are required, namely : (I.) the *assembly drawing* (see Plate 16), which shows the relative positions of the parts in the completed structure, together with its most general dimensions ; and (II.) the *detail drawings* (see Plate 15), which give the form, arrangement, and dimensions of the parts of an object taken separately.

If a drawing is made from an existing object, the data consists of sketches and measurements from the object.   When a new design is required to be expressed, the drawing is usually worked out from explanatory sketches, calculations, and previous drawings.   The making of drawings involving data of the latter kind, belongs to some one of the various branches of engineering or architectural construction, and therefore lies outside the scope of this book.

In the duplication of objects, requiring no engineering or architectural experience, the following practice should be adopted.

61. Sketching the Object. (See also Art. 50.)   Sketch only such views of an object, or a part, and only as much of each view, as are necessary to make the working drawing.   Supplement the freehand sketching by the use of straight-edge and compass whenever this will save time.   Cultivate the habit of rendering neatly and legibly all sketches and dimensions, as frequently in office practice one man makes the sketches, while another makes the drawing from the sketches.

62. Measuring the Object. (See also Art. 50.)   Be careful not to omit any essential measurements ; if the object is not near by, an omission may mean both trouble and expense.

(*a*)   Small, nicely machined pieces should be measured with a micrometer caliper.   For ordinary work a two-foot rule and machinist's calipers are sufficient. Whenever practicable, take as the base lines for measurements finished edges (as

---

* The drawings reproduced in Plates 12 to 16 were made primarily to illustrate arrangement and to give practice in reading and rendering. Though in each case the object might readily be made from them, they are not, in a strictly technical sense, working drawings, since their form, as will presently be seen, is not wholly in accordance with office practice.

f, f, Fig. 135) of the object, and, if possible, take all measurements from the
same base lines (compare Figs. 135 and 136).

(*b*) *Circular holes.* In locating a hole, do not attempt to measure to its
center, but take measurements as shown in the following examples. Place on the

sketch the measurements just as they are taken, although
in many cases this is not the way they should appear on
the working drawing (see *e*, Art. 65).

A single hole. (Fig. 111.) From edge *A*, measure
distances *AB* and *BC*. When making the working drawing, find the position of the center of the hole by adding
to the distance *AB*, one-half of the diameter *BC*.

Fig. 111.

Equally spaced holes in a straight line. (Fig. 112.)
Lay the rule along the center line *AB*, measure the distance between corresponding edges, as *A* and *B*, of two holes at a considerable distance apart, and state the

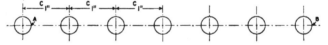

Fig. 112.

number of spaces between the holes in question — in this case 6. To find the
distance between the centers of adjacent holes, when making the working drawing, divide *AB* by the number of spaces between the holes.

Equally spaced holes in a circle, as, for example, a series of bolt holes. When
the number of holes is *even* (Fig. 113), measure between corresponding points, as
A and *B*, in opposite holes. Re-
cord the number of holes ; if necessary, one hole may be located
with respect to the vertical or
the horizontal axis of the piece.
In making the working drawing,
draw the circle of centers (bolt
circle) with a radius equal to one-
half of *AB*, and space the circle
for the required number of holes.

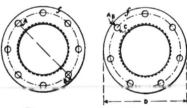

Fig. 113.   Fig. 114.

When the number of holes is *odd* (Fig. 114), the measurements must be taken
from the inner or outer edge of the piece, according to which is the smoother. In
the case shown in Fig. 114 the inner edge of the piece is threaded, and therefore
the holes are located from the outer edge. Measure the outside diameter *D*, the
distance *AB*, and the diameter *BC* of a hole. When making the working draw-

ing, draw the bolt circle with a radius equal to one-half of D, minus the sum of AB and one-half of BC, and space the circle for the required number of holes.

If the holes are not equally spaced, each hole may need to be located separately. No general rule can be given.

63. Making the Drawing ; General Directions. Ordinarily, a working drawing may be said to include the layout and preliminary drawing, a tracing, and a blue print. Both the tracing and the blue print are called working drawings. The tracing is usually retained in the office, while the blue print is sent away for shop or other outside use.

(a) *The layout* should be made directly on duplex detail or a similar paper. After the scale of the drawing, the number and arrangement of the views, and the spaces for dimensions and title have been decided upon, the pencil drawing for the tracing should be continued on the layout. (See Art. 52.)

(b) *The tracing.* (See also Art. 52.) For most working drawings of machine construction the line widths D, $E — J$, Fig. 55, are appropriate. In drawings for steel construction, since many of the lines come close together, it may be necessary to make the lines somewhat narrower, both for clearness and to prevent them from running together. It must be borne in mind, however, that, if the lines on the tracing are made much narrower than lines $D — J$, Fig. 55, they are very likely to be weak in the blue print, and hence, except for dimension and extension lines, they are unsatisfactory.

(c) *The blue prints.* The process of obtaining copies by blue printing is explained in Chapter XI.

64. Planning the Drawing. (a) *The scale.* In order that a working drawing may best serve its purpose, it is necessary, when planning it, to take into account not only the immediate facts concerning the making of the drawing as such, but also the facts connected with its ultimate use. Thus, for example, if the drawing is to be used at the bench or lathe, it should not be of a size which will be unwieldy, or which cannot be easily scanned by the mechanic. On the other hand, the drawing must be of a sufficiently large scale to enable the workman to read easily all of its parts and dimensions.

(b) *The number of views.* In the interest of both convenience and economy the number of sheets should be as small as possible ; that is, as many views of an object should be placed on the same sheet as there is room for without overcrowding. *Do not give views, or parts of views, which are unnecessary.* The views selected should be such as will best set forth the essential characteristics of the object. Thus, the interior of an object is usually represented best by one or more *sectional views* rather than by dotted lines, which are more likely to confuse than to explain. For example, compare the projection, Fig. 115, having all invisible lines shown, with the corresponding projection, Plate 16, in which most of the invisible lines are omitted.

(*c*) *The arrangement of the views.* In planning an arrangement, endeavor to anticipate the probable space that will be required for dimensions placed outside of

Avoid

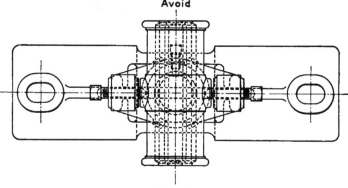

Fig. 115.

the views, so that dimensions belonging to different views shall not conflict. (See Art. 65.) In all drawings of machinery place the views according to *third angle projection.** This gives the following arrangement : —

Top view.

Left-hand side or end view.   *Front view.*   *Right-hand side or end view.*

Bottom view.

* The system of projection most generally adopted in mechanical engineering is that of the third angle of descriptive geometry, although first angle projection is also followed to some extent. In architecture and civil engineering, first angle projection appears to be generally adopted; that is, the top view is placed below the front view, the left-hand side view to the right of the front view, and the right-hand side view to the left of the front view.

That two systems of projection are used in practice is unfortunate, for several reasons. When a person becomes accustomed to one angle, it is more or less confusing to read drawings made in the other angle. Then, besides this inconvenience, there is always the chance for costly mistakes in the shop, resulting from reading the drawing in the angle other than the one intended by the draftsman. To illustrate, take a case which occurred just at the time of writing. A former student in architecture, having patented an electric street railway signal, made a drawing from which to have the device manufactured. Naturally, as an architect, he made his drawing in the first angle. When the lot of castings was delivered, he found that certain parts which should be on particular faces of certain pieces had been made on opposite faces, and that the castings were useless. The pattern makers read his drawing in the third angle instead of in the first; hence the mistakes.

65. Dimensioning. (*a*) *Necessary dimensions.* In order to dimension a working drawing properly, it is evidently necessary to know what dimensions should be given. This requires not only the ability to discriminate between necessary and unnecesary measurements, but also some practical knowledge of shop construction, in order to know when, and what, particular measurements may be needed by the mechanic. As a simple example, take the case of a hexagonal bolt head. In forging the hexagonal figure of the head, the mechanic can easily work with reference to its short diameter, or "distance between the flats"; whereas, if the measurement of the diagonal (long diameter), or distance between corners, is given, he must figure out the short diameter, or work at a disadvantage. As a further illustration, if a drilled hole is dimensioned according to the directions for measuring (Art. 62), that is, by giving its diameter and the distance to the edge of the hole, the workman must figure for the center, since he must know at what point to set the point of the drill.

If unfamiliar with shop requirements, it is best, when dimensioning a working drawing, to give *all essential measurements used in making the drawing.*

(*b*) *Forms of dimensions.* The general form of a dimension — which includes the numerals expressing the measurement, the dimension and extension lines, and the arrow heads — is described in Art. 42.

Because of the general use in shop work of the two-foot rule, dimensions less than two feet should be given in inches; if greater than two feet, in feet and inches.

(*c*) *General system of placing dimensions.* No dimension should appear upside down, when a drawing is read from the bottom or the right-hand side. A satisfactory system of *placing* is illustrated by the several positions of the diameter dimension shown in Fig. 116, which will be found convenient for reference in doubtful cases. It will be seen that, in this system, all dimensions read from the bottom of the sheet, except those on the vertical line *AB*, which read from the right-hand side of the sheet.

Fig. 116.

(*d*) *Position of dimensions; clearness.* It is always important to place a dimension in such a position that it may readily be seen; its connection with the part of the drawing to which it refers must at once be evident; and it must not conflict with or obscure the drawing. To secure these results may require considerable ingenuity and judgment, as varying conditions must be met in different ways, but always in conformity with the following practice, which must be regarded, by the beginner, as invariable.

On assembly drawings give only the most general dimensions, such as overall dimensions and distances between centers. Do not give unnecessary dimensions nor repeat a dimension on the same drawing, and do not leave any calculating,

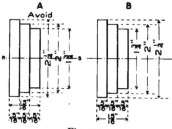

Fig. 117.

however simple, to be done by the workman. Do not fail to give the totals (overall measurements) of intermediate measurements.

Whenever a dimension tends to obscure the drawing, or if the actual place of measurement will be more clearly shown thereby, place the dimension outside of the view by means of extension lines (*B*, Fig. 117); do not, however, place the dimension so far from the place to which it refers that

it shall appear detached. In the example of placing, Fig. 118, the distance, *a*, between dimensions, and between a dimension and an edge of the drawing, is made

equal to the distance, *b*, used for the height of the numerator and the denominator. It should be understood, however, that the distances *a* and *b* are not to be measured; all placing should be determined by eye.

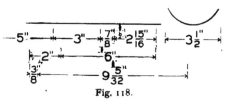

Fig. 118.

Extension lines must always be drawn parallel (see *error*, Fig. 119), and at right angles (*B*, Fig. 120) to the direction in which the measurement is taken.

Avoid

Fig. 119.

(See *error*, *A*, Fig. 120.) A dimension line must always be drawn parallel to the line of measurement.

Never place a dimension on a line of the drawing, on a center line (*A*, Fig. 117), on an extension line, or on a dimension line of another dimension (*B*, Fig. 121).

Place overall dimensions outside of detail dimensions, and place a longer dimension outside of a

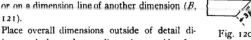

Fig. 120.

shorter one (*B*, Fig. 117). If a long measurement is placed between the drawing and a shorter measurement (*A*, Fig. 117), it is necessary to cross the dimension line of the longer measurement by the extension lines of the shorter, an arrangement which may cause confusion.

In some cases greater clearness may be obtained by *staggering* the dimensions. This consists, *A*, Fig. 122, in breaking each continuous line of dimensions (compare *A* with *B*, Fig. 122) or in breaking up symmetrical columns of dimensions, as the diameters in Fig. 117, by placing the dimensions in a diagonal column, or by placing alternate dimensions in separate columns. *Note.* As the system *A*, Fig. 122, is not in common use, it should not be adopted unless authorized.

Fig. 121.

(*e*) *Special forms of dimensions.* When the space in which a dimension should be placed is too small to take the dimension without interference or crowding, the following forms may be used: —

If space permits, place the arrow heads as usual, but place the dimension at

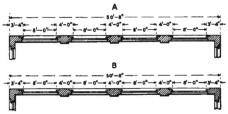

Fig. 122.

Fig. 123.

one side and connected with the dimension line by a narrow freehand line (see the left-hand example, *A*, Fig. 123). When there is no room for the arrow heads, reverse them and place the dimension and freehand line as in the preceding case (see middle example, *A*, Fig. 123). Or reverse the arrow heads and place the dimension in line with the arrow heads, omitting or using dimension lines extending outward from the points of the arrows according to preference (see the right-hand example, *A*, Fig. 123). A combination of the last two forms is shown in *B*, Fig. 123.

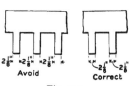

Fig. 124.

A comparison of confused and clear dimensioning is given in Fig. 124. The dimensions given in *A*, Fig. 124, are intended for the widths of the projections, but are so placed that they give the widths of the spaces between the projections — a result due to the position of the measurement relatively to the arrow heads, and the omission of a freehand line drawn from the numerals to the arrow heads. Compare the confused arrangement, *A*, with the clear one, *B*, Fig. 124.

Radii of circles. If there is sufficient room, a radius measurement should be placed between the arc and its center, Fig. 125, and the center enclosed in a free-

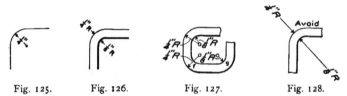

Fig. 125.　　　　Fig. 126.　　　　Fig. 127.　　　　Fig. 128.

hand circle of about $\frac{1}{8}$" diameter. If there is not room for this, the center should be ignored, the dimensions placed according to Fig. 126, and the letter "*R*," or "*Rad,*" invariably added. In the case of concentric arcs, the larger radius may be reversed (Fig. 126), or either of the forms shown in Fig. 127 may be used. (The second example includes the radii *f* and *g*.) Do not distort the form, Fig. 126, by placing the quantities too far from the arrow heads (see Fig. 128).

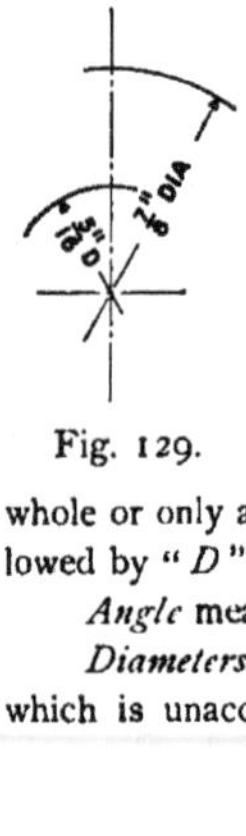

Do not give the radii of non-essential curves such as those representing corners rounded for a finish.

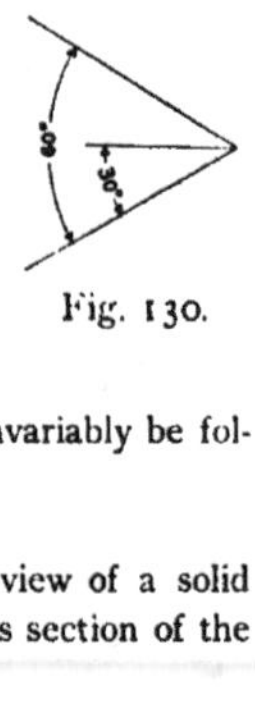

Diameters of circles. When a diameter dimension is given with one arrow head and only a portion of its dimension line, Fig. 129, whether the whole or only a part of the circle be shown, the dimension must invariably be followed by "*D*" or "*Dia.*"

Fig. 129.

Fig. 130.

Angle measurements should be given as shown in Fig. 130.

Diameters of solids. When a diameter is given on a side view of a solid which is unaccompanied by a view showing the shape of the cross section of the

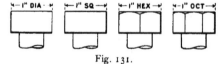

Fig. 131.

solid, the dimension should include an abbreviation descriptive of the cross section, as follows: "*D.*" or "*Dia.*" (diameter); "*Sq.*" (square); "*Hex.*" (hexagonal); "*Oct.*" (octagonal). (See Fig. 131.)

In a composite view, where a partial section is combined with an outside view, the dimension line on one side of a dimension may be drawn of indefinite length and its arrow head omitted, Fig. 132, which shows that the measurement reads to a point corresponding to that at which the one arrow head is placed.

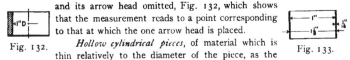

Fig. 132.

Fig. 133.

Hollow cylindrical pieces, of material which is thin relatively to the diameter of the piece, as the shell of an engine boiler, should be dimensioned according to Fig. 133; that is, give the thickness of the material and both the inside and the outside diameters of the piece.

Circular holes. A hole must be dimensioned by giving its diameter (*B*, Fig. 134) and the distance of its center from a finished edge of the piece (*A*, Fig. 134). The dimensions taken in measuring the object (*b*, Art. 62) must be reduced to bring them to this form. When there are several holes, their centers should be located as shown in Fig. 135, using the same two edges of the piece as base lines (see *error*, Fig. 136).

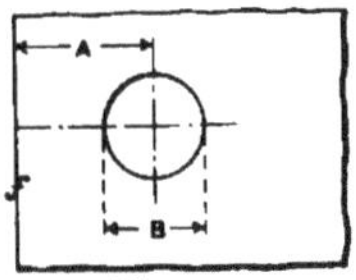

Fig. 134.

Holes in a straight line. When the holes are all of the same diameter and equally spaced, they may be dimensioned as follows: Locate the centers of the first and last holes in the line, and give the overall dimension between these centers. At each end of the line, dimension, two or three times, the distance

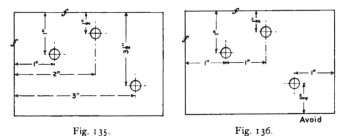

Fig. 135.

Fig. 136.

from center to center of adjacent holes, also two or three diameters of the holes; but omit the intermediate dimensions — the omission will indicate that the spacing and diameters, respectively, are uniform throughout.

Holes in a circle. When holes are equally spaced around a circle, give the diameter of the circle passing through the centers of the holes, as shown at *A*,

Figs. 137 and 138, or as shown at *B*, Fig. 138. Calculate the diameter according to *b*, Art. 62. Give also the diameter of one or two of the holes ; this will indicate that the others are of the same size. No other dimensions are usually necessary,

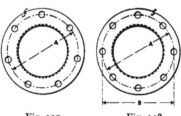

Fig. 137.   Fig. 138.

since equal spacing is understood unless otherwise stated, and the number of holes, unless very numerous, may be counted from the drawing.

When holes in a circle are irregularly spaced, or not all of the same size, all necessary measurements must be given.

66. Conventions. (*a*) *Line conventions* are given in Fig. 55.

When a sectional and an outside view are combined (composite view or projection), the two views should be separated by a dash-and-dot line (see Plate 15).

(*b*)   *Shade lines*, if indicated, should be treated according to Art. 58.

(*c*)   *Shading* is rarely seen in working drawings. If it is used, the light,

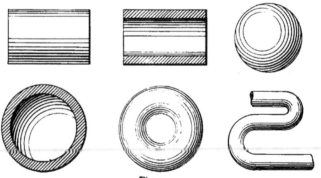

Fig. 139.

rapidly rendered examples shown in Fig. 139 should be taken as a guide, rather than the examples given in Fig. *D*, Plate 4.

(*d*)   *Materials*. The conventions shown on Plate 4 may be used (see Art. 38). The graining, Fig. *A*, Plate 4, should be used sparingly, and only in cases where wood might be mistaken for a metal.

(*e*) *Finished surfaces.* When a surface of an object is required to be finished in the shop, the fact is indicated by means of a lower-case italic *f* (Fig. 140). The letter should never be placed on a surface, as view *B*, Fig. 140, but it should be placed on a line, straight or curved, which is an edge view of the surface. Thus the three *f*'s in *A*, Fig. 140, indicate that the front, top, and back surfaces of the block are to be finished, since the three lines on which the letters are placed are edge views of the surfaces in question. Let the cross stroke of the *f* intersect the line on which the letter is placed; if the line is horizontal, the cross stroke should incline at 45°. When all the surfaces of an object are to be finished, the *f* should not be used, but instead place on the drawing the note, "*Finish all over.*"

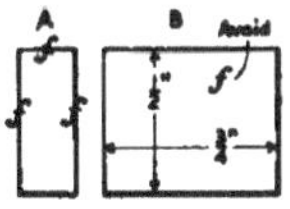

Fig. 140.

(*f*) *Screws and tapped holes.* Any of the conventions given in Figs. 98 and 99 is appropriate. In addition to the convention the number of threads to an inch, as "16 THD." (*A*, Fig. 141), must always be stated. To save time, the inclination and spacing of the lines of a screw thread convention should be determined by eye; care should be taken that the opposite ends of the longer lines shall have the same relative positions as the points of a thread, Fig. 94, but do not attempt to represent the *number* of threads to an inch.

Additional conventions are shown in Fig. 141. The dotted lines, *A*, indicate that a piece is threaded for the distance covered by the lines in question. The convention *B* is used to represent a longitudinal projection of an invisible threaded or tapped hole; the same convention is also used for the screw when in position. The dotted circle, *C*, shows that the hole is threaded; if the hole is invisible, both circles should be dotted. In the last two cases, besides the thread convention, always give the diameter of the tap and the number of threads to an inch, as "¾" — 16 TAP" (*B* and *C*, Fig. 141), unless already given in the *immediate vicinity*. At *D* is shown a section through the tapped hole which receives the screw *A*; it should be noticed that the lines of the thread convention are drawn in the opposite direction to those of convention *A*.

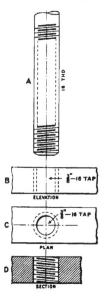

Fig. 141.

When a section is taken through a screw and the tapped hole into which it fits, none of the thread conventions can be used; it is therefore necessary to draw the sections of the threads (see Fig. 142).

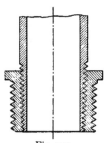

Fig. 142.

67. Letters and Numerals. Plain letters and numerals, such as those given in Plate 5, should be used. They must be perfectly legible, and for speed, if under $\frac{1}{4}''$ in height, should be stroke rendered (Art. 40). All letters and numerals should be sufficiently large to be easily read by the workman, but they should not be more conspicuous than the drawings.

As vertical numerals take less space laterally than inclined ones, if there is any probability of crowding, only vertical numerals should be used. Comparing A and B, Fig. 143, it will be seen that the dimensions in B are the clearer; even when the size of the inclined numerals is materially reduced, they still appear more crowded in the narrower spaces than the vertical (compare B and C).

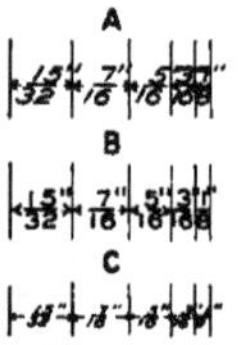

Do not omit the dash between feet and inches, Fig. 144. Do not make straight top 3's; if carelessly made, they may be taken for 5's, Fig. 144. Do not make open-top 4's; if carelessly made, they may be taken for 7's. Always make the dividing line of a fraction horizontal; if inclined, it may lead to an error in reading: thus, for example, the $1\frac{1}{16}''$ (Fig. 144) might easily be taken for $\frac{11}{16}''$.

A

B

C

Fig. 143.

68. Titles. Office drawings are usually filed away in drawers; hence, for easy reference, *the title must be placed in the lower right-hand corner of the drawing.* An expert draftsman can render excellent letters up to $\frac{3}{8}''$ in height by means of the ruling pen sharpened for lettering (c, Art. 40). Several pens should be kept, each sharpened for different line widths. In larger letters the edges of the outlines

Avoid   Avoid

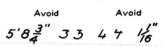

Fig. 144.

should be stroke rendered, the spaces filled in with the pen, and the letters rapidly finished freehand.

CHAPTER IX.

69. Pseudo-pictorial Representation. This form of drawing includes two general systems : *axonometric* projection, which includes *isometric* projection ; and *oblique* projection, which includes *cavalier* or *cabinet* projection. In either of these systems an object is represented by a single view having a more or less pictorial effect — a fact which makes these methods occasionally convenient in describing a construction to persons not familiar with representation by plan and elevation.

In *axonometric* projection an object is supposed to be projected on a plane, usually vertical, by projectors at right angles to the plane ; its special case of *isometric* projection results from a particular position of the object. In *oblique* projection an object is supposed to be projected on a plane, usually vertical, by parallel projectors not at right angles to the plane ; its special case of *cavalier* or *cabinet* projection results when the projectors make an angle of 45° with the plane.

Isometric drawing is a practical modification of *isometric projection*, and because of its greater simplicity and more general usefulness will be considered first.

70. Isometric Drawing. All rectangular objects are bounded by three systems of parallel edges. In isometric drawing, lines belonging to these systems are drawn parallel, respectively, to three lines, as HK, HJ, and HD, Fig. *B*, Plate 17, known as the *isometric axes*. One axis is usually taken vertical, the other two at 30° with the horizontal.

Isometric drawings are usually made either from the actual object or from plans and elevations. Let Fig. *A*, Plate 17, represent the plan (C) and elevation (B) of a cube supposed to be viewed in the direction of the arrow S. It will be seen that, in the isometric (Fig. *B*) all edges of the object which make 45° with the vertical plane X (Fig. *A*) are drawn parallel and at 30° with the horizontal, and that the vertical edges are drawn vertical. The lengths of the edges in the isometric drawing are made equal to the edges in the object, as shown by the measurements given on both drawings.

71. Isometric of Rectangular Solids. (*a*) *A cube* (*Figs. A and B, Plate 17*). *Construction:* From any assumed point, *D*, Fig. *B*, draw, of indefinite lengths, the vertical line *DH*, and the lines *DE* and *DF*, making angles of 30° with the horizontal. From point D lay off on DH, DE, and DF the measurement ($\frac{3}{4}''$) given in Fig. A. Draw, at 30° with the horizontal, the back edges EG and FG of the base ; put in the vertical edges, and draw the top of the cube.

(*b*) *An object composed of rectangular solids.* In Fig. C is given the plan (B) and elevation (A) of a model. Consider, as in the projections of the cube (Fig. A), that the object is placed so that its horizontal edges make 45° with the imaginary vertical plane X, Fig. C. *Construction :* Starting at any assumed point, D, Fig. D, and reading all measurements from Fig. C, draw the part FGE. A projecting piece, as dbg, must be built up from its surface of contact ; that is, from the surface, $abcF$, common to both pieces. *The left-hand piece :* Starting at the corner F, Fig. D, draw the surface of contact $Fabc$, its edges equal to $Fabc$, Fig. C; draw the edges ad, be, cg, and Ff, and connect for the end $degf$. *The right-hand piece :* Lay off Gm, Fig. D, equal to Gm ($\frac{1}{8}''$), Fig. C. Draw the surface of contact $mkhj$, Fig. D, its edges equal to $mkhj$, Fig. C. Draw the vertical edges and connect for the top.

(*c*) *A rectangular recess.* The method of drawing a recess is illustrated by the mortise in the piece D, Figs. J and K. The construction is begun with the rectangle $oprq$, and completed as shown. (Compare the similarly lettered lines in Figs. J and K.)

72. Non-isometric Lines. Curves and straight lines oblique to either of the three systems of a rectangular object must be located by means of rectangular co-ordinates taken parallel to two (or all three) of the systems. Let A and B, Fig. E, be two vertical faces of a cube, and C a horizontal face, upon which are drawn oblique straight lines and curves as shown. The method by which these oblique lines and curves would be drawn in isometric is illustrated in Fig. G.

(*a*) *The lines M and N, Fig. E.* As the line M intersects the two edges, KH and FD, of the cube, make Ka and Fb, Fig. G, equal to the distances Ka and Fb, Fig. F, and connect the points a and b. *The line N.* The extremity d of the line, Fig. E, is located by the co-ordinates Hc and cd; the extremity f, by He and ef. The isometric of these co-ordinates gives the position of the points d and f, Fig. G.

(*b*) *Polygons.* The isometric should be made from a preliminary drawing showing the true shape of the given polygon. Circumscribe a rectangle, as $kmqp$, C, Fig. E, about the polygon ; draw the isometric of the circumscribing rectangle, and locate the corners of the polygon (see Fig. G).

(*c*) *Plane curves.* In drawing the isometric of a curve, first make a drawing showing the true shape of the curve, and draw rectangular co-ordinates locating a number of its points. In the case of a circle it is convenient to inscribe and circumscribe parallel squares, the sides of which establish the co-ordinates of eight points in the circle ; namely, at the middle points of the outer square and at the corners of the inner square.

Construction of the vertical circle, B, Fig. E. Draw the diagonals and diameters of the face *HJED*, Fig. *G*. On a diameter, as *OP*, lay off the distance, as *Od*, Fig. *E*, between the inscribed and circumscribed squares. Construct the inscribed square, and draw the curve as indicated by the lettering in Figs. *E* and *G*.

The horizontal circle, C, Fig. E. The construction is the same as the preceding, as indicated by the corresponding letters in Figs. *E* and *G*.

(*d*) *Planes and solids.* The faces of the cube, and the lines drawn on them, Fig. *E*, are repeated in Fig. *F*. It will be seen, however, that by means of the additional projections *D*, *E*, and *G*, the lines drawn on the faces, Fig. *E*, become in Fig. *F* the projections of planes and solids. An isometric drawing of the planes (*A*), cylinders (*B* and *C*), and the prism (*C*) would be begun by making a drawing precisely the same as Fig. *G*. The lines and figures drawn on the surface of the cube, Fig. *G*, are the lines and surfaces of contact of the planes and solids, Fig. *F*, which would be completed in the isometric as follows : —

The planes, A, Fig. F. The edges *a'g'*, *h'j'*, *d'k'*, and *f'm'* of the planes *M* and *N* are perpendicular to the surface of the cube, and hence are represented in the isometric, Fig. *H*, by lines drawn at 30° and equal to the lengths shown in *D*, Fig. *F*.

The cylinders. Each cylinder is supposed to be circumscribed by a square prism ; the base of the prism circumscribed about the horizontal cylinder (*B*, Fig. *F*) is represented by the isometric square, *acrp*, Fig. *H*. From the corner *a* of the square, Fig. *H*, draw the edge *aj* of the prism equal to the height, *g't'*, of the cylinder, Fig. *F*. Draw the isometric of the outer base of the prism ; inscribe the isometric circles, and draw the elements of the cylinder tangent to them. The construction is similar for the vertical cylinder, *C*, Fig. *F*.

The inclined brace F, Fig. K. The inclined edges of the brace should be obtained by connecting the ends of the brace (surfaces of contact), which are drawn first because they are bounded by isometric lines.

(*e*) *Spheres ; double curved surfaces of revolution.* Make a preliminary drawing, showing the true size and figure of the given solid. Take a series of parallel sections, and find the isometric of each. A line tangent to these sections is the required curve. An isometric drawing of a sphere is a circle, but its diameter is about two-ninths greater than the diameter of the given sphere.

(*f*) *An approximate method for drawing the isometric of a circle,* by the use of circular arcs, is given in Fig. 145. As all the oblique lines are drawn at either 30° or 60°, the construction should become clear by inspection of the figure. Turning the figure so as to bring either diameter, 2—2, vertical, will give the correct positions of the construction lines for circles lying in vertical planes. This method is usually sufficiently accurate, except in the case of very large circles.

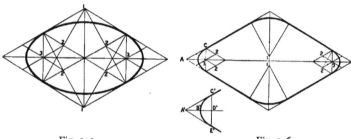

Fig. 145.   Fig. 146.

The application of the method to rounded corners is shown in Fig. 146. When AC does not exceed $\frac{1}{4}$ inch, however, the construction at A becomes so small that it is better to proceed as follows : Make distance $A'B'$, Fig. 146, equal to distance $C'D'$, and sketch a freehand curve through points C', B', and E'.

73. Shade Lines in Isometric Drawing. The light is supposed to take the direction V, Fig. B, Plate 17, parallel to the diagonal KE of the cube; the shadow of line ab is represented by the line ba', parallel to the diagonal KJ. The shade lines of the cube comprise the boundary $LJHDF$; the practical shade lines of an isometric drawing are those which can be determined by eye. Another method is to shade all lower and right-hand (sharp) edges (see c, Art. 58).

74. Applications of Isometric Drawing. In Figs. J—P, Plate 17, are shown practical uses of isometric drawing. The detail, Fig. M, is drawn from its plan and elevation A and B, Fig. L. The detail of framing, Figs. O and P, is taken from an old-time working drawing of a bridge across the Saco River at Hiram, Me. While the type of construction is obsolete, the drawing suggests the value of isometric in representing a construction not easily shown in plan and elevation (see A and B, Fig. N).

Perhaps the most useful and general application of isometric is to describe details of building construction (see Plate 18).

75. Axonometric Drawing. Isometric drawing is a special case of axonometric drawing, resulting from a particular choice of the *axes* (see Art. 70) taken to represent the three systems of parallel edges of a rectangular object. The general case of axonometric drawing is illustrated by the cube shown in Fig. 147. The lines AB, AC, and AD are the *axonometric axes*. One axis, AD, is usually

Plate 17

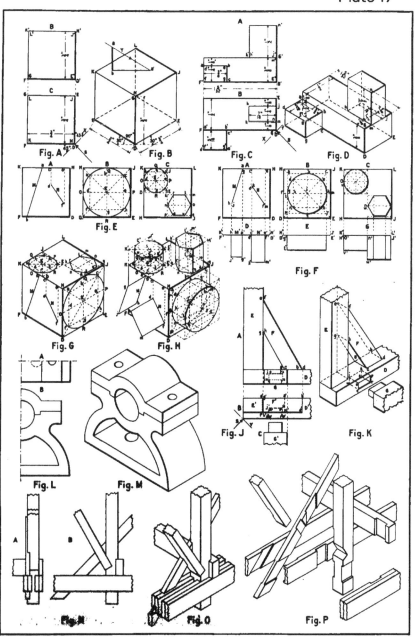

Fig. A Fig. B Fig. C Fig. D
Fig. E Fig. F
Fig. G Fig. H
Fig. J Fig. K
Fig. L Fig. M
Fig. N Fig. O Fig. P

drawn vertical, and made equal to the true length of the edge of the object; the other axes, AB and AC, are drawn at any convenient angles, and their lengths determined by descriptive geometry. In general, the lengths of AB and AC will

not be equal to the true lengths of the lines of the object, as in isometric drawing, but will be proportional to these lengths; these proportions, when determined, are known as the *scales of the axes*. For example, if AC were found to equal one-half the length of the line in the object, then all lines in the system parallel to AC must be drawn one-half of their true lengths.

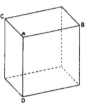

For practical purposes the axes may be arbitrarily assumed in such positions, and of such lengths, as will give a desirable representation. The following system agrees closely with the results derived by descriptive geom-

Fig. 147.

etry, and gives a satisfactory appearance. In Fig. 147 draw AD vertical, AB at 12° with the horizontal, and AC at 45° with the horizontal. Make AD and AB each equal to the lines of the object, and AC equal to two-thirds the length of the line of the object, or in these proportions: for example, if AD and AB are laid off from the scale of $1\frac{1}{2}'' = 1$ foot, then AC should be laid off from the scale of $1'' = 1$ ft. For drawing the lines at 12° with the horizontal, a special triangle is convenient.

Curves, and straight lines not parallel to the axonometric axes must be determined by means of co-ordinates parallel to the axes. The methods are similar to those of isometric drawing, except that the scales of the axes must always be applied to the co-ordinates.

76. Oblique Projection. A cube drawn in oblique projection is shown in Figs. 148 and 149. In this system one axis, AD, is taken vertical, another axis,

Fig. 148.

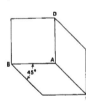

Fig. 149.

Fig. 150.

AB, is taken horizontal, and both these axes are laid off in their true lengths. The third axis, AC, may be drawn at any oblique angle, and laid off to any desired proportion of its true length. In Figs. 148 and 149 the line AC is made equal to its true length, giving the special case of oblique projection known as *cavalier* or *cabinet* projection.

An idea of the general effect obtained by oblique projection may be gained from Fig. 150. (Compare with the same object drawn in isometric, Fig. *M*, Plate 17.) It will be seen that the drawing, Fig. 150, shows the true shape of the front and back faces of the object. The lines connecting these faces are drawn at 45°, and their lengths are laid off to a scale one-half of that used for the front and back faces.

Oblique projection may occasionally be useful; but, as the principles and methods are similar to those of isometric drawing, further explanation is unnecessary.

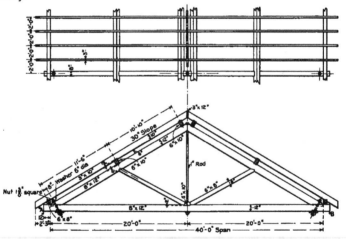

Fig. 151.

STUDY PLATE 15.

Isometric drawing.

Use Whatman's hot-pressed paper. The size of the sheet is to be 10" x 14", with a ruled border line 9" x 13". Make a finished isometric drawing, in ink, scale $\frac{1}{4}'' =$ 1 ft., of the roof truss, rafters, etc., given in Fig. 151. Place the point B, Fig. 151, $\frac{3}{4}''$ from the lower, and 3" from the right-hand line of the border; draw the line AB upward to the left.

Plate 18

Pl·14·

Detail · of · Framing ·

Scale ¾ inch = 1 foot.

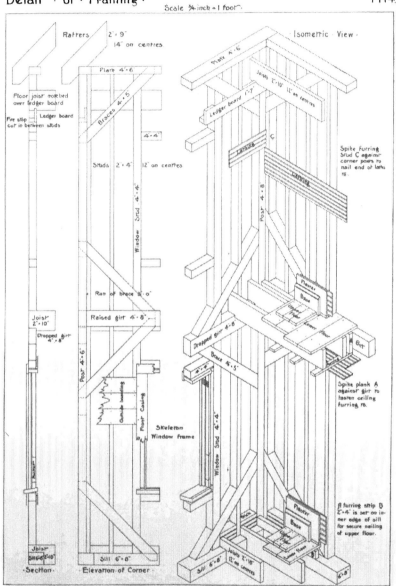

Rafters 2" × 9"
14" on centres

Plate 4" × 6"

Braces 4" × 5"

4" × 4"

Studs 2" × 4" 12" on centres

Window Stud 4" × 4"

Floor joist notched over ledger board

Fire stop cut in between studs

Ledger board

Run of brace 3'·0"

Raised girt 4" × 8"

Joist 2" × 10"

Dropped girt 4" × 8"

Post 4" × 6"

Outside sheathing

Floor Casing

Skeleton Window frame

Joist 2" × 10"

Sill 6" × 8"

·Section·

Sill 6" × 8"

·Elevation of Corner·

Isometric · View ·

Plate 4" × 6"

Joists 2" × 10" 12" on centres

Ledger board 1" × 7"

C

Post 4" × 8"

Spike furring stud C against corner posts to nail end of laths rs.

Plaster
Base

Dropped girt 4" × 8"

Brace 4" × 5"

A Girt

Spike plank A against girt to fasten ceiling furring rs.

Window Stud 4" × 4"

A furring strip B 2" × 4" is set on inner edge of sill for secure nailing of upper floor.

Plaster
Base

Joists 2" × 10" 12" on centres

Sill 6" × 8"

6" × 8"

Omit invisible lines, and the bolts. Put in shade lines. Do not give the dimensions (Fig. 151).

Design and letter on the drawing the title "Roof Truss, Scale $\frac{1}{4}$" = 1 ft."; also letter "Plate 15," your name and the date; all of the letters are to be drawn.

STUDY PLATE 16.

Isometric drawing.

Use Whatman's hot-pressed paper. The size of the sheet is to be 10" x 14", with a ruled border line 9" x 13". Make an isometric drawing, in ink, of an assembly of the clamp, Fig. 87. Take with the dividers the distances directly from Plates 13 and 14.

Before beginning the drawing, make a layout to determine the position of the drawing on the sheet. Show the bar F, and the glued pieces C, C, C, C, Fig. 87. Break the bar and the piece K, in order that each shall fall within the ruled border line. Represent the thread of the screw according to convention C, Fig. 98, but draw the lines of the convention curved, and parallel to the edge of the hole through the part B, Fig. 87.*

In the binder, G, first draw it as if square in cross section, and then shape it freehand. Treat the knobs on the part K according to e, Art. 72. Omit all invisible edges; put in shade lines.

Design and letter on the drawing the title "Cabinet Maker's Clamp"; letter also "Plate 16," your name and the date; all of the letters are to be drawn.

* A template may be made having one edge straight, and the curve so inclined that the template may be moved along the edge of the T-square.

CHAPTER X.

77. A Wash is a mixture of India ink — or a water color — and water, applied with a brush.

(*a*) *The materials for wash drawing.* Apart from the considerable skill necessary to manage washes successfully, the quality of a result depends largely upon the quality of the materials used. For a brief course of study, intended merely to give the practice necessary for a fairly presentable coloring of small areas — as on maps and surveyor's plans — the materials given on page 2 may be used.

If, however, the subject of wash drawing is approached with view to elaborate rendering — as in architectural drawing, see Plate 22 — only the best materials should be used. Get the best quality of Japanese ink; select a stick which dissolves readily under the moistened finger, and in which the dissolved ink tends to a brown shade rather than to a blue. The brown sable brushes made by Winsor and Newton, while expensive, are the best; the red sable of the same make are the next best. Satisfactory sizes are the Nos. 6 and 11, and No. 4 of the extra large series. In purchasing a brush, selection should be made from half a dozen or more after trying each, as follows: Thoroughly saturate the brush; then, with only enough water to bring the hairs to a point, test the brush by drawing it briskly downward and across the finger or the edge of a water glass. Select the brush in which the hairs show the most life, and, in springing to place, form the sharpest and straightest point.

If a wash is applied to a free or unstretched light-weight paper (as 72 lb. Whatman), it causes the paper to expand; and, when dry, the paper will be more or less uneven or wrinkled. Hence there should be used a very heavy paper (not lighter than 140 lb. Whatman); a Whatman or a water-color paper mounted on cardboard; or any of these papers made into a water-color block. The best results are obtained on stretched paper.

(*b*) *To stretch paper.* Place the paper on the drawing board, with the better side of the paper uppermost. Rule a pencil line parallel to and 1 inch from each edge of the sheet. At each corner of the paper, outside of the ruled line, cut out a triangular slip of the paper, as shown at *C*, *D*, *E*, and *F*, Fig. 152 (*A*). Lay a straight-edge along each pencil line, and bend the paper until it stands perpendicular to the drawing board. With a sponge well filled with water, moisten the surface, *L*, taking care to keep the upturned margin *dry* and not to abrade the paper. When the paper has become quite limp, apply to the outer surface of the margin, *G'*, *H'*, *J'*, *K'* (*B*), a liquid glue, strong paste, or mucilage ; * turn down the margin, *A*, Fig. 152, and at the same time pull each portion of the margin, *G*, *H*, *J*, and *K*, toward the edge of the board.

Fig. 152.

If the stretching is properly done, the paper will dry out perfectly flat. It is necessary, however, to watch the work until the adhesive has set. If the paper dries out faster than the adhesive, the contraction of the paper will cause the glued margin to slide away from the edges of the board, and the paper will dry slack ; hence, if the adhesive dries but slowly, the pulling out of the glued margin should be repeated, and the paper again moistened, if necessary. On the other hand, if the paper is very wet and is pulled too tight, the contraction may cause the paper to tear, or may warp or even split the drawing board.

(*c*) *To prepare an India ink wash.* Never use stale ink, but grind it fresh at each exercise. The prepared waterproof liquid inks are wholly unsuitable for wash drawing. As the ink should be free from sediment, it is best to grind it on a china slab. See that the slab and the nest of saucers are clean and free from dust. Grind a reasonable quantity of ink — which need not be so thick as for line drawing — to be used as a first supply in mixing the washes. In order that the ink may be as " smooth " as possible, bear down but lightly on the stick, and, before mixing a wash, let the ink stand a few minutes, so that the sediment may settle. If it is required that a wash shall match a given wash or printed area (Plate 20), the wash must be tried on spare paper from time to time, when preparing it ; the shade or value must not be judged until the wash is thoroughly dry. To mix a wash, half fill three or four of the saucers with clear water ; add to the water in each saucer a varying quantity of ink, taken from the slab, so as to produce several washes of different values. The final wash should then be obtained by modifying any one of the preceding washes by admixture from the others, or by addition of clear water, until the required value is obtained. It is difficult to obtain a required light value by a direct mixture of dark ink and water.

78. A Flat Wash. A wash is said to be *flat* when it dries out to a uniform value, and is free from a clouded appearance, streaks, and spots. A *direct* wash signifies one obtained by a single application of the liquid wash. A *built up* wash is obtained by repeated washes.

---

* The liquid glue is the best, but it softens very slowly when the glued paper is soaked in cleaning the board.

As a means of stating, very roughly, the amount of wash a brush should contain for different results, let the expressions a "full brush" mean one holding as much of the wash as is possible without dripping, and a "dry brush" indicate one holding the least amount which will permit the brush to transmit wash to the paper.

(*a*) *To lay a direct flat wash on an unbroken surface.* With an H to 3H pencil, rule lightly a boundary for the wash, as *KADM*, Fig. 153. Have ready a

generous supply of wash and also some blotting paper. Incline the drawing at an angle of from 10° to 15° with the horizontal. With the *largest* brush half-full, starting near the upper right-hand corner, *C*, Fig. 153, and carrying the brush firmly and accurately along the ruled line, *DA*, lay a broad strip of the wash, *C* to *B*. The inclination of the paper causes the wash to settle or form a *pool* (a

Fig. 153.

portion of which is shown at *GH*) along the lower edge of the strip, and the management of this pool ("flowing the wash") is the principal consideration in laying flat washes. Working very rapidly, and using a full brush, so as to keep the pool as full as possible without overflowing, guide the pool from its first position, *FH*, to a second position, a portion of which is shown at *KL* ; this should be done with strokes of the brush perpendicular to the line *FH*. The pool must now be left for

an instant in order to carry the wash, before it has begun to dry, accurately into the corners *A* and *D*, and to the ruled border at *AE* and *DJ* ; for this use a dry brush, either a small one, or the one already in use, dried on the blotting paper. After the edges of the wash have thus been attended to, the pool must be advanced with the full brush, the edges again managed with the dry brush, and the process repeated until the bottom ruled line is reached. As this line is

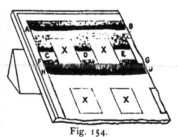

Fig. 154.

approached, the wash should be gradually exhausted from the brush and the finishing strokes should be made with the dry brush.

(b)   *To lay a flat wash on a broken surface.*   Let Fig. 154 represent a surface broken up by the rectangles *X—X*, which are not to receive the wash.   The process is the same as that described in the preceding paragraph, except that the single pool, as *AB*, is broken up, as it advances, into three pools, as at *C*, *D*, and *E*, when the wash must be treated as three simultaneous washes.   Along the line, *FG*, of the lower edges of the rectangles, the pools *C*, *D*, and *E* are again joined to form a single pool, as *HJ*.   With the attention divided between the alternate use of the wet and the dry brush, the keeping of all advancing edges wet, and the accurate following of the ruled lines, it is evident that considerable dexterity is necessary, and that the draftsman must work rapidly.

(c)   *Precautions necessary to secure a flat wash.*   In order that a wash shall dry out flat, it is necessary that the value of the wash shall not vary; therefore, as a wash quickly settles in the saucer, it must be remembered to stir up the wash each time it is applied to the drawing.   Moreover, it is necessary that the wash shall be fed uniformly into the pool; that is, in floating a wash, the brush must hold the same amount each time it is applied to the paper.   The results of a non-observance of the latter requirement, and of working too slowly, are shown in Fig. *F*, Plate 19.   In making the drawing for this cut, the wash lay flat while handled at a proper speed; but on slower working the slight cloudiness, at *DE*, appeared.   At *FG* the wash was allowed to become almost dry, hence the streak.   The streak at *HJ* resulted from an excessive pool at *KL*, which caused the wash to flow back into the partially dried portion between *FG* and *KL*.   The spot at the corner *N* resulted from carrying the pool into the corner, instead of exhausting the wash, as described above.

A wash tends to dry out in a hard edge, as at *BC* and *CE*; and hardness is increased if the wash collects in a groove made by using too hard a pencil or by bearing down on the pencil.   Hard edges may be avoided, as at *GJ*, by slightly thinning the wash, at its very edge, with clear water, used in a dry brush.   A soft edge may also be obtained by building up a wash with two washes, and keeping the boundary of the second wash — without the aid of a ruled line — a trifle inside the boundary of the first (see the edges *AB* and *AD*).

Before laying a wash, see that the paper is clean; if at all soiled, it should be washed with clean water applied with a soft sponge.   The use of a rubber is likely to injure the surface.   A direct wash is likely to work better if the surface is first gone over with a very light wash of the India ink, or with water containing a trace of yellow ochre.   A wash is more likely to be flat if the paper can be kept slightly damp (not moist).

In laying in a narrow stripe (see Plate 20), the wash should be managed as described for a flat wash, but the pool should contain less of the wash.   For very large areas, a foot square or more, a camel-hair sky brush may be used, but the edges should be managed with the sable brush.

79. Graded Values. The light and shade of a surface may be expressed by means of a series of graded flat washes, or by grading a direct wash.

(*a*) *To lay a wash of uniform gradation, by means of flat washes.* Let it be required to obtain a gradation with four washes, as shown by *A—D*, Fig. *A*, Plate 19. Rule, lightly, pencil lines representing the edges of the several washes, *D*, Fig. *A*. Repeat these lines several times on spare paper to be used in the trials necessary to determine the value of the dry washes. Prepare a wash which shall give the lightest value, Fig. *A ;* then a second wash which, when applied over the first one, will give the next value, Fig. *B ;* and so on. Lay the washes in the order of the values, beginning with the lightest, as shown in *A—D*, Fig. *A*. The edges will be somewhat softer if the washes are laid in the reverse order ; that is, beginning with the darkest value, and overlapping the subsequent washes. *Each wash should be thoroughly dry before the next one is applied.*

The modelling of a cylinder by using four flat washes is shown in *A—D*, Fig. *B*, Plate 19. The shading of a sphere, by the same method, is given in *B—E*, Fig. *C ;* the outlines for the washes may be drawn as indicated in *A*.

A water line (see Plate 21) should be begun with the lightest wash, *A*, Fig. *D*, Plate 19. As the width of the wash must not vary, the edges may merge in the narrow places, thus giving a continuous wash, as between the shore lines on the right-hand side of *A*. In the second wash, *B*, Fig. *D*, the edges merge only between the island and the shore line.

The drawing of the hook, Fig. *G*, was built up with a considerable number of flat washes applied without the aid of pencil outlines.

(*b*) *To lay a graded, direct wash.* This gradation, *E*, Fig. *A*, Plate 19, may be effected by beginning the wash with the darkest value, and then gradually dilut-ing the pool of wash, as it advances, by adding clear water. A better way is to prepare several washes of graded values ; begin with the darkest, and lighten the advancing pool with the prepared washes taken in the order of their values. Clear water should also be at hand to modify the advancing pool, if necessary.

(*c*) *A built-up graded wash.* Proceed as in the previous paragraph, but use much lighter washes. In *H*, Fig. *B*, is shown a cylinder shaded by this method ; the several stages are indicated in *E*, *F*, and *G*.

80. Methods of Correcting a Wash Drawing. (*a*) *Washing.* Minor spots and streaks may be *partially* eliminated by washing with clean water applied with the brush. The part should be thoroughly soaked and scrubbed *with the brush*, and the water then removed with blotting paper ; in medium to light washes a

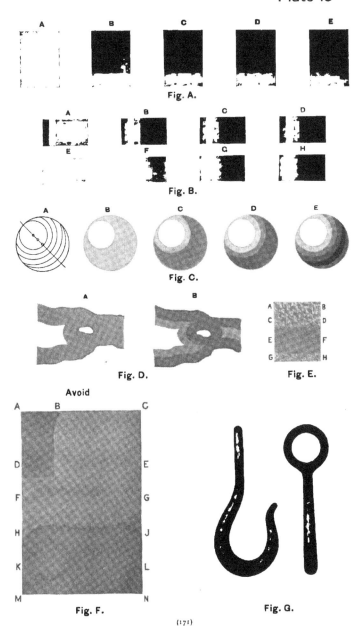

Plate 19

Fig. A.

Fig. B.

Fig. C.

Fig. D.

Fig. E.

Avoid

Fig. F.

Fig. G.

result may not be perceptible until the process has been repeated several times. In order to make radical corrections, it is necessary to sponge the whole drawing, although a part of the drawing, if isolated by clear paper, may be so treated. All traces of the wash must be removed from the paper by repeated applications of *clean* water, as the slightest discoloration of the water will stain the paper. If the whole drawing is to be washed, it may be placed directly under the water faucet. The sponge must be a soft one, and used lightly, so as not to rub up the surface of the paper.

(*b*) *Stippling.* This is the modifying of a dry wash by the placing of small spots of liquid wash applied with the point of a dry brush. To illustrate the process, let it be required to build up a value *EF*, Fig. *E*, Plate 19, wholly by stippling. The paper is first spotted with the point of the dry brush, as shown at *AB*. The spots of clear paper at *AB* are broken up by additional touches, as at *CD*, but without overlapping the first spots. The clear paper is further eliminated by a third spotting, as at *GH*, with smaller touches of the wash, and the process continued until a required value, *EF*, is produced. Stippling consumes much time, and considerable practice is necessary in order to judge correctly the value of the liquid wash used in the spotting.

Stippling is permissible only when it may be the means of saving considerable time in redrawing. Brush washing with pure water and stippling may be alternated.

81. Miscellaneous Notes. (*a*) *Ink lines.* In a wash drawing proper — such as the architectural drawing, Plate 22 — all edges which would be represented by lines in an outline drawing are represented by the value differences of the washes. Such edges must never be further defined by ruled lines in black ink, although an edge may be accented, occasionally, by a line of *wash* applied with the ruling pen. In elaborate engineering drawings, maps, and plans, on which flat washes are used, the inking should follow the laying of the washes ; but, in working drawings and rush work, washes may be applied after the inking, providing the lines are drawn with waterproof ink.

Chinese white, a pigment of value in making pictorial illustrations, should not be used on washes in architectural or engineering drawing.

An ink line should never be drawn around the edge of a cast shadow.

(*b*) *Care of the brushes.* The brushes should be washed out thoroughly after use, the hairs brought to a straight point, and the brushes kept in a brush holder. If a brush is laid in a drawer, care must be taken that the point does not come in contact with anything ; a point which dries bent is usually spoiled. India ink ground black for line work should never be used in a wash brush, as it is difficult to remove all traces of the ink.

STUDY PLATE 17.

Flat and graded washes.

Use Whatman's cold-pressed paper. The size of the finished plate is to be 10″ x 14″, with a ruled border line 8″ x 12″. Stretch the paper (*b*, Art. 77), and rule lightly with an H pencil the boundaries of the washes, Plate 20, according to the measurements given in Fig. 155. Any erasure in connection with the penciling should be done lightly with a velvet rubber, so as not to injure the surface of the paper.

It is required to lay flat and graded India ink washes (Arts. 78 and 79) of the values shown on the plate. Take the areas in the order of their ar-

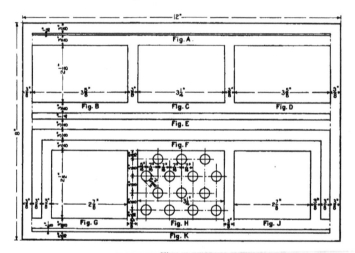

Fig. 155.

rangement. The values, Figs. *A*, *B*, *E*, *F*, *H*, and *K*, must be obtained with a single wash ; not more than two washes may be used for Fig. *D*, and not more than three washes for Fig. *C;* the number of washes for Figs. *G* and *J* is not limited. Stippling must not be used.

Ink only the ruled border line of the plate, and the lettering "Plate 17," your name, and the date (drawn letters).

Plate 20
(Study Plate 17)

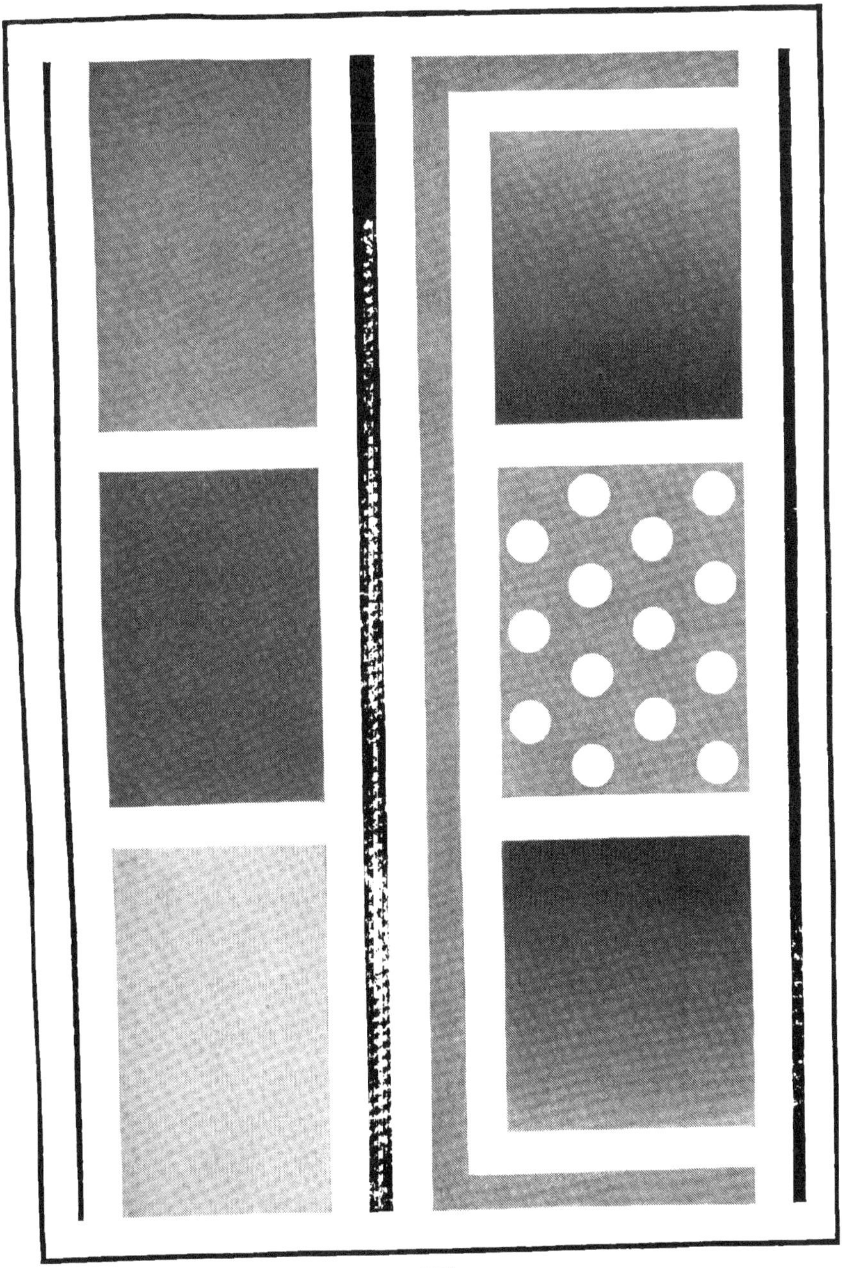

STUDY PLATE 18.

Modeling by graded washes.

Use Whatman's cold-pressed paper, stretched. The finished plate is to be 10″ x 14″, with a ruled border line 8″ x 12″.

It is required to arrange and shade six figures similar to *B*, *D*, *F*, *H*, *J*, and *L*, Fig. 156 (also see Plate 21), changing the sizes and the proportions so as to fill the sheet satisfactorily. Make a layout for the arrangement, and then lightly draw the figures with an H pencil. Model the forms with graded washes; take Plate 21 as a general guide, but keep the darkest value in each figure lighter than that on the plate.

Ink only the ruled border line of the plate, and the lettering "Plate 18," your name, and the date (drawn letters).

STUDY PLATE 19.

Modeling by flat washes.

Proceed as in Study Plate 18, but let the modeling be done with graded *flat* washes (see Figs. *A*, *C*, *E*, *G*, and *K*).

STUDY PLATE 20.

Modeling and water lines by graded washes.

Use Whatman's cold-pressed paper, stretched. The finished plate is to be 10″ x 14″, with a ruled border line 8″ x 12″.

It is required to arrange and shade three figures similar to *M*, *O*, and *Q*, Fig. 156 (also see Plate 21), changing the sizes and proportions so as to fill the sheet satisfactorily. Make a layout for the arrangement, and then lightly draw the figures with an H pencil. Render the pipe and the water lines with graded washes; take Plate 21 as a general guide, but keep the darkest value in each figure lighter than that on the plate.

Ink only the ruled border line of the plate, and the lettering "Plate 20," your name, and the date (drawn letters).

STUDY PLATE 21.

Modeling and water lines by flat washes.

Proceed as in Study Plate 20, but let the pipe and the water lines be rendered in *flat* instead of graded washes (see Figs. *N* and *P*).

STUDY PLATE 22.

Modeling and water lines by flat and graded washes.

Use Whatman's cold-pressed paper. The size of the finished plate is to be 10″ x 14″, with a ruled border line 8″ x 12″. Stretch the paper, and lay out the figures, Plate 21, according to the measurements given in Fig. 156.

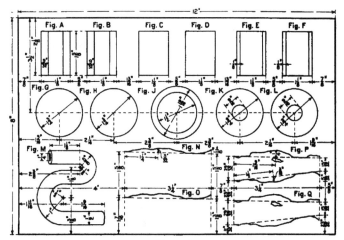

Fig. 156.

It is required to make an accurate copy of Plate 21, for values, and treatment (see Arts. 78 and 79).

Ink only the ruled border line of the plate, and the lettering "Plate 22," your name, and the date (drawn letters).

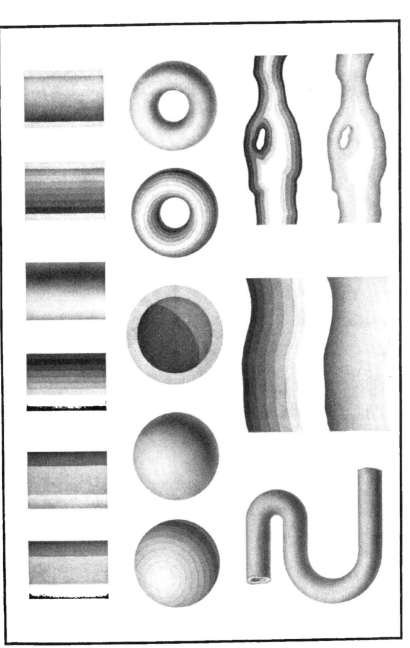

Plate 21
(Study Plate 22)

Plate 22

FROM THE WASH DRAWING BY A. TOURNAIRE.
The size of the original is 24 x 38 inches.

CHAPTER XI.

82. Mechanical Copying. (*a*) *A tracing-paper transfer.* Place tracing paper over the outline required to be copied. Trace the outline; turn the paper over; go over the traced line with a soft pencil, and rub down the lead. Reverse the tracing paper, place it on the drawing and go over the traced line with a 3H pencil. Instead of rubbing lead on the back of the traced line, a transfer paper — prepared by rubbing powdered lead uniformly on tissue paper — may be placed between the tracing and the drawing paper.

(*b*) *A celluloid transfer.* A sharper and narrower line than by the preceding methods may be obtained by making the tracing on thin celluloid, with a sharp steel point. Go over the back of the traced line with a steel point, and then rub into the line the lead of a black or a blue pencil. Dust off the superfluous lead, place the celluloid with the lead-filled line next to the paper, and then rub the line briskly with the burnisher.

(*c*) *A rubbing.* The outlines of forms in relief, such as ornament and lettering, may be obtained, if their edges are fairly sharp, by placing thin paper over the forms, and then rubbing over their edges.

83. Enlargement and Reduction. (*a*) *By pantograph.* Irregular figures, such as maps and diagrams, may be enlarged or reduced with a fair degree of accuracy by means of an inexpensive pantograph. An expensive form of this instrument is necessary for accurate engineering work.

(*b*) *By triangulation, base lines, and offsets.* First establish the principal points of the given outline by a series of triangles, built up from a base line connecting any two important points in the given outline, and then tie in additional points by means of offsets from the sides of the triangles taken as additional base lines. Begin the enlargement or reduction by drawing a line proportional to the principal base line in the original, and, starting from this base line, draw triangles and offsets proportional to those on the original. Having thus determined the position of the principal points, the outline must be sketched in by eye. The proportional distances used in constructing the triangles may be obtained by the use of two scales, or with the proportional dividers. If the size of the original and of the reduction or enlargement will permit, place the paper on which the copy is to be made close to the original, and obtain the triangles for the copy by means of lines — drawn by sliding the triangle (Art. 23) — parallel to the sides of the triangles in the original.

Fig. 157.

(c) *By proportional squares.* Let it be supposed that it is required to make from a photograph or a cut, Fig. 157, a large wall diagram to illustrate this machine.

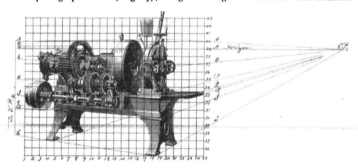

Fig. 158.

Circumscribe a rectangle about the photograph, and rule lines, as 1, 2,— 43, Fig. 158, dividing the rectangle into squares. Begin the diagram, Fig. 159, by drawing a rectangle of the size required, and having the sides proportional to those of the rectangle, Fig. 158. Divide the sides 1 — 25 and 25 — 43, Fig. 159, into the number of parts used on the original, Fig. 158, and draw the horizontal and vertical lines as shown. Find the vanishing points VP_1 and VP_2 (the latter is at the intersection of the lines A, B,—L, Fig. 159; also see note on the cut of the original, Fig. 158) by producing several of the convergent lines, and draw the horizon.

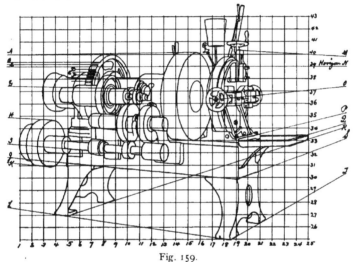

Fig. 159.

Draw the horizon, Fig. 159, and locate the vanishing points by laying off on the horizon the necessary (proportional) distances. Note the point where a line crosses the edge of a square in the original, Fig. 158, and sketch the line through the corresponding point (judged by eye) in the corresponding square on the diagram, Fig. 159. The convergent lines should be ruled with the aid of a long straight-edge passed through the proper vanishing point.

If there is objection to drawing directly lines on an original, the squares may be drawn on tracing paper placed over the original.

In making a copy directly from a flat (actual) object, it is sometimes convenient, first, to obtain its outline by running a pencil around the object laid on paper, after which the outline may be enlarged or reduced by any of the methods described.

84. The Blue-print Process. Blue-print paper is a white paper with a coating which is sensitive to light. So long as the paper is protected from light, the coating can be easily removed by washing; but on exposure to sunlight the coating turns blue, and becomes insoluble in water.

A print is obtained by means of a wooden frame set with glass, and having a removable back lined with felt. The drawing, preferably a tracing, is placed with the ink lines next to the glass; the blue-print paper with the sensitized surface next to the tracing. The frame is then exposed to direct sunlight, which, passing through the portion of the tracing cloth not covered by the ink lines, acts on the sensitized surface, while the portions of this surface protected by the lines of the tracing are not affected. After an exposure varying from twenty seconds to five minutes, dependent upon the "speed" of the paper, the print is thoroughly washed in a tank of running water or by means of a hose. The print is then hung up, that it may drain properly and dry flat. A good print shows clear white lines on a uniform blue ground.

When the back of the printing frame is in position, the blue-print paper, the tracing, and the glass should be in close contact. If the contact is imperfect, it should be corrected by means of a felt pad; otherwise more or less light will pass under the lines, with the result that edges of the white lines in the print will be blurred, or the lines tinged with blue.

If a print is under-exposed, it washes out to a pale greenish blue, giving a weak contrast with the white lines. If a print is over-exposed, the light is likely to work through the lines of the tracing, in which case the lines on the print wash out to a pale blue instead of white, and the ground becomes too dark or turns gray.

Special care must be taken to exclude all light from quick papers — a covered case or can is convenient — and to open them only in a subdued light or a dark room. Quick papers may be printed by electric light.

As excellent prepared blue-print papers, of different speeds, can be purchased at a moderate price, it is usually not worth while for the draftsman to coat his own paper. The process, however, is as follows : —

For a paper requiring an exposure of about five minutes, mix separately,

(1) Red Prussiate of Potash (recrystalized)	1 part	(by weight)
Water	5 parts "	"
(2) Citrate of Iron and Ammonia	1 part "	"
Water	5 parts "	"

Working in a subdued light, mix equal parts of these solutions, and apply to the paper with a sponge, or a flat 3-inch camel-hair brush. (For a quicker paper more of the citrate of iron solution must be used, but the color of the print will not be so good.) A good quality, hard surface paper should be used. As the solutions *1* and *2* are not affected by light, they may be kept on hand ready for use.

Blue lines on a white ground (positive prints). Make a preliminary negative on thin brown-print paper — obtained of dealers in drawing supplies — with the tracing reversed ; that is, with the ink lines next to the sensitized surface. Substitute the brown-print negative, which shows white lines on a brown ground, for the tracing, and proceed as described for the prints on a blue ground.

Good blue prints can be made from drawings on thin, or on bond paper. When a blue print is to be made from a drawing on thick paper, the drawing may be made more transparent by wetting with naphtha or gasolene, which dries out without injury to the drawing. For a sharp print it is necessary, in order to exclude the oblique rays which will pass under the lines, to lay the drawing face down, but with the disadvantage that the print is reversed.

When a part of a drawing is not wanted in a print, the part may be covered with thick paper placed between the tracing and the glass of the printing frame. Pencil lines and spots may be removed from a tracing by sponging with naphtha or gasolene, which does not injure the cloth or the ink lines.

For making alterations on a blue print, white lines may be made with soda, or any alkali, dissolved in water, with a small quantity of gum arabic added to prevent the mixture from spreading on the blue print ; or black drawing ink may be used on light blue prints, and red ink on dark ones.

Blue prints required to stand much handling, if the size permits, may be mounted on binder's board and then shellacked.

85. Process Drawing. A drawing made to be reproduced by any of the photo-mechanical processes, such as photo-engraving and photo-lithography, is termed a *process drawing.*

The practical bearing of the special requirements in this kind of drawing will be better understood from a brief description of the processes of photo-engraving.

86. Line Plates. A line plate is one which reproduces an outline drawing.

The photographic negative. A negative of the drawing is made by means of a wet plate, so prepared that the film or negative can be pulled ("stripped") from the glass. After exposure in the camera, the negative is developed, subjected to various chemical processes, and then stripped.

The zinc transfer. When dry, the film is reversed, placed on a thick plate of glass, and a print is made on a sheet of highly polished zinc or copper, coated with a mixture of glue sensitized with bichromate of ammonia. The mounted film and the sheet of metal are placed in a heavy printing frame, and exposed to light, as in blue printing ; the light causes the sensitized glue to become insoluble in water. The print is washed, to remove the portions of the coating not affected by the light, and then "burned in" by exposure to an intense heat, which carbonizes the insoluble glue remaining on the plate.

Etching. The plate is placed in a solution of nitric acid, which eats out the surface not covered by the carbonized glue. After a short exposure to the acid ("first bite") the plate is dried and brushed over with powdered dragons-blood, to protect the sides of the lines from the acid, which would otherwise attack the lines beneath the carbonized surface, and thus produce what is technically called "undercutting." The etching and the application of the dragons-blood are done three times (four times for extra deep etching).

Finishing. The plate now goes to the engraver, to be cleaned up by hand; after which the relief of the lines is increased by cutting the ground of the plate deeper with the "routing" machine. The plate is then proved on a printing press; finally it is nailed to a block to bring it to the height of printer's type.

87. Drawings for Line Plates. The first step toward securing excellence in a line plate is attention to details in making the original drawing.

For greater sharpness of line in the plate, and that imperfections in the drawing may be eliminated as far as possible, the drawing should be made larger than the plate. Over-reduction, however, may result in broken or ragged lines in the plate, a weak-looking cut, or one so small that its details may not easily be seen. The best results are obtained from a drawing made from one-third to one-half larger than the plate. Make the drawing on smooth, white Bristol board,* and keep it clean. See that *all* lines of the drawing have sharp, smooth edges, and that the lines are black. If the plate is to be considerably smaller than the drawing, ample allowance must be made for the reduction in line widths, in order that the lines on the plate may not come so fine that, unless retouched on the zinc — not always skilfully done — there will be danger of the lines breaking down from undercutting. In "forcing" the line widths in a drawing, as necessary for a considerable reduction, the beginner must not be misled by the difference between the appearance of his process drawing and that of an engineering drawing, since, as compared with the latter, the former may appear altogether too heavy. In dimensioning and lettering a drawing, it is specially important to allow for the reduction, as a reduction satisfactory for the drawing may bring the letters to a size not easily read. A reducing glass will give some idea of the appearance of a reduction, the approximate size of which may be obtained by measuring the image on the glass.

When very narrow lines on the plate are desired, it should be remembered that ragged lines tend to thicken in the photographic negative. A perfectly smooth line on the drawing will give a narrower line on the plate than will a ragged line of the same or even less width.

A rubber must be used with caution on a process drawing, as its use will quickly make black lines gray.

* Drawings on tracing cloth give excellent results, but correction is more difficult.

Special methods. White lines — ruled or freehand — on a black ground may be obtained by laying in the ground and then drawing the lines in Chinese white. The white (Plate 3) must be diluted with water, and the proper thickness must be judged from trial lines, which must be allowed to dry, as the dry pigment is whiter than the liquid. The black ground must be laid in with waterproof ink.

Dash lines may be ruled solid in waterproof ink, and the dashes may be obtained by cutting the line with the white, applied with a brush.

Methods of correction. Minor corrections can be made to advantage with Chinese white. For example, the width of a line may be reduced and ragged edges removed by ruling a line with the white, applied with the ruling pen. Letters may be cleaned up and their outlines corrected by means of the white, applied with a small, sharp-pointed brush. It is necessary, however, to use the white with caution ; as the pigment accumulates rapidly, several applications of it may be sufficient to cast a shadow when the drawing is photographed. If a correction is not satisfactory, scrape off the white and correct again. As a slight film of the white over black will prevent the black from photographing, it is a good plan to look over all Chinese white corrections with a magnifying glass before sending the drawing away. Lines should never be cleaned up with a steel eraser, as furred edges in the drawing are exaggerated by the camera.

When necessary to redraw a line, it is better to draw the line on thin, smooth paper pasted over the line than on a surface furred up by erasure. The edges of a patch, however, cast a shadow which must be removed by the engraver ; hence the edges of the patch must be kept as far away as possible from lines of the drawing, in order that the engraver may have sufficient space in which to work.

If any considerable portion of a drawing must be corrected, the place may be patched as just described, or the place may be cut out and fresh piece of cardboard inserted. If the latter method is adopted, the inserted cardboard must match that on which the drawing is made, since a difference in the *shade* and cleanliness of the two pieces may show in the negative, and thereby affect the quality of the line plate.

88. Half-tone Plates. The reproduction, by a plate, of light and shade drawings must be done by the half-tone process, which is similar to that of the line plate with the following exceptions : The drawing is photographed through a screen — composed of lines ruled on glass — placed close to the sensitized plate. The image of the screen appears on the negative, breaking up the image of the drawing into minute points of varying character ; and the duplication of these points on copper constitutes the half-tone plate. The etching of the plate — done with perchloride of iron instead of with nitric acid — takes considerably more time than the etching of a line plate.

The coarseness of the screen is measured by the number of its lines to an inch, and the particular screen used depends upon the character of the printing in which the plate will be used. For the coarsest newspaper work a screen of 65 lines is used, and for fine book work a screen of 200 lines, although as high as 400 lines has been used. (The screen for Plate 22 was one of 175 lines.) The finer finish of half-tone plates is the work of skilled (hand) engravers, and finishers who re-etch locally with a brush. A plate is darkened by burnishing and lightened by re-etching.

89. Drawings for Half-tone Plates. Black and white drawings for half-tone plates are made in the usual manner. Lamp black is perhaps the best medium to use, although good results attend the use of India ink, charcoal gray, etc. Unless it is desired to have the grain of the paper show in a print, very smooth paper must be used. The whitest of paper is reproduced by a tint in the plate (see Fig. *G*, Plate 19, showing the tint of the paper on which the hook was drawn) ; hence, for a reproduction on a white ground, the tint must be removed by the engraver and the routing machine (see Plate 22). When the tint is to be removed, the engraver will be materially assisted, in mechanical subjects, if the boundary of the drawing is defined by a line in Chinese white.

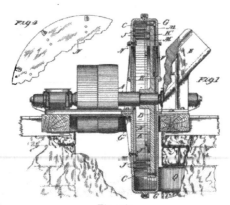

Fig. 160.

90. Patent Office Drawing. An application for a patent must be accompanied by a drawing made in accordance with the requirements contained in the extract from the "Rules of Practice of the United States Patent Office" which follows. Any system of drawing, such as orthographic projection, perspective, isometric, etc., may be used, and the one selected should be that which will show the invention to the best advantage.

The cuts, Figs. *A* and *B*, Plate 23, reproduced from the " Rules of Practice," indicate the character of patent-office drawings. It will be seen that the upper left-hand view, Fig. *B*, is in isometric, while the other views are orthographic projections. In the case of a drawing such as is shown in Fig. *A*, considerable line shading is permissible, but it should be used sparingly in drawings of machinery. For example, the drawing, Fig. 160, is overshaded ; other objections to this drawing are the placing of letters, as *N, K*, and *Q*, on shaded surfaces and the unnecessary representations of wood and stone.

A plate giving the conventions to be used in drawings of electrical apparatus will be found opposite page 86 of the " Rules of Practice."

Extract from the " Rules of Practice of the United States Patent Office " :

". . . The applicant for a patent is required by law to furnish a drawing of his invention whenever the nature of the case admits of it.

. . . The drawing may be signed by the inventor, or the name of the inventor may be signed on the drawing by his attorney in fact, and must be attested by two witnesses. The drawing must show every feature of the invention covered by the claims, and the figures should be consecutively numbered if possible. When the invention consists of an improvement on an old machine the drawing must exhibit, in one or more views, the invention itself, disconnected from the old structure, and also in another view so much only of the old structure as will suffice to show the connection of the invention therewith.

. . . Three several editions of patent drawings are printed and published,— one for office use, certified copies, etc., of the size and character of those attached to patents, the work being about 6 by $9\frac{1}{2}$ inches ; one reduced to half that scale, or one-fourth the surface, of which four are printed on a page to illustrate the volumes distributed to the courts ; and one reduction — to about the same scale — of a selected portion of each drawing for the Official Gazette.

This work is done by the photolithographic process, and therefore the character of each original drawing must be brought as nearly as possible to a uniform standard of excellence, suited to the requirements of the process, and calculated to give the best results, in the interests of inventors, of the office, and of the public. The following rules will therefore be rigidly enforced, and any departure from them will be certain to cause delay in the examination of an application for letters patent :

(1) Drawings must be made upon pure white paper of a thickness corresponding to three-sheet Bristol board. The surface of the paper must be calendered and smooth. India ink alone must be used, to secure perfectly black and solid lines,

(2) The size of a sheet on which a drawing is made must be exactly 10 by 15 inches. One inch from its edges a single marginal line is to be drawn, leaving the "sight" precisely 8 by 13 inches. Within this margin all work and signatures must be included. One of the shorter sides of the sheet is regarded as its top, and, measuring downwardly from the marginal line, a space of not less than $1\frac{1}{4}$ inches is to be left blank for the heading of title, name, number, and date.

(3) All drawings must be made with the pen only. Every line and letter (signatures included) must be absolutely black. This direction applies to all lines, however fine, to shading, and to lines representing cut surfaces in sectional views. All lines must be clean, sharp, and solid, and they must not be too fine or crowded. Surface shading, when used, should be open. Sectional shading should be made by oblique parallel lines, which may be about one-twentieth of an inch apart. Solid black should not be used for sectional or surface shading.

(4) Drawings should be made with the fewest lines possible consistent with clearness. By the observance of this rule the effectiveness of the work after reduction will be much increased. Shading (except on sectional views) should be used only on convex and concave surfaces, where it should be used sparingly, and may even there be dispensed with if the drawing is otherwise well executed. The plane upon which a sectional view is taken should be indicated on the general view by a broken or dotted line. Heavy lines on the shade sides of objects should be used, except where they tend to thicken the work and obscure letters of reference. The light is always supposed to come from the upper left-hand corner at an angle of forty-five degrees. Imitations of wood or surface graining should not be attempted.

(5) The scale to which a drawing is made ought to be large enough to show the mechanism without crowding, and two or more sheets should be used if one does not give sufficient room to accomplish this end ; but the number of sheets must never be more than is absolutely necessary.

(6) The different views should be consecutively numbered. Letters and figures of reference must be carefully formed. They should, if possible, measure at least one-eighth of an inch in height, so that they may bear reduction to one twenty-fourth of an inch; and they may be much larger when there is sufficient room. They must be so placed in the close and complex parts of drawings as not to interfere with a thorough comprehension of the same, and therefore should rarely cross or mingle with the lines. When necessarily grouped around a certain part, they should be placed at a little distance, where there is available space, and connected by short broken lines

Plate 23

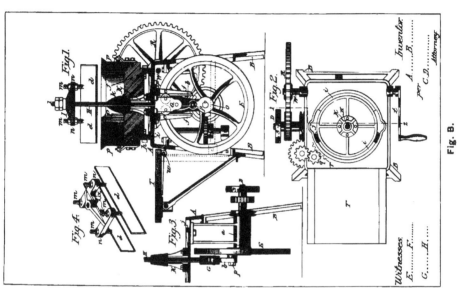

Fig. B.

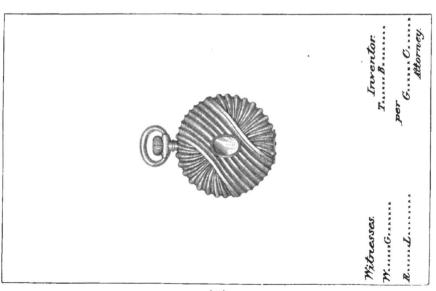

Fig. A.

with the parts to which they refer.   They must never appear upon shaded
surfaces, and, when it is difficult to avoid this, a blank space must be left in
the shading where the letter occurs, so that it shall appear perfectly distinct
and separate from the work.   If the same part of an invention appear in more
than one view of the drawing, it must always be represented by the same
character, and the same character must never be used to designate different
parts.

(7)   The signature of the inventor should be placed at the lower right-
hand corner of each sheet, and the signatures of the witnesses at the lower
left-hand corner, all within the marginal line, but in no instance should they
trespass upon the drawings.   (See specimen drawing. . . .)   The title should
be written with pencil on the back of the sheet.   The permanent names and
title will be supplied subsequently by the office in uniform style.

When views are longer than the width of the sheet, the sheet should be
turned on its side and the heading will be placed at the right and the signatures at the left, occupying the same space and position as in the upright
views, and being horizontal when the sheet is held in an upright position ; and
all views on the same sheet must stand in the same direction.   One figure
must not be placed upon another or within the outline of another.

(8)   As a rule, one view only of each invention can be shown in the
Gazette illustrations.   The selection of that portion of a drawing best calculated to explain the nature of the specific improvement would be facilitated
and the final result improved by the judicious execution of a figure with
express reference to the Gazette, but which might at the same time serve as
one of the figures referred to in the specification.   For this purpose the figure
may be a plan, elevation, section, or perspective view, according to the judgment of the draftsman.   It must not cover a space exceeding 16 square
inches.   All its parts should be especially open and distinct, with very little
or no shading, and it must illustrate the invention claimed only, to the exclusion of all other details.   (See specimen drawing.)   When well executed, it
will be used without curtailment or change, but any excessive fineness, or
crowding, or unnecessary elaborateness of detail will necessitate its exclusion
from the Gazette.

(9)   Drawings should be rolled for transmission to the office, not
folded."

DECIMAL EQUIVALENTS OF FRACTIONS OF AN INCH.

$\frac{1}{64}$	.0156	$\frac{9}{64}$	.1406	$\frac{17}{64}$	.2656	$\frac{25}{64}$	.3906	$\frac{33}{64}$	.5156	$\frac{41}{64}$	.6406	$\frac{49}{64}$	.7656	$\frac{57}{64}$	.8906
$\frac{1}{32}$	.0313	$\frac{5}{32}$	.1563	$\frac{9}{32}$	.2813	$\frac{13}{32}$	.4063	$\frac{17}{32}$	.5313	$\frac{21}{32}$	.6563	$\frac{25}{32}$	.7813	$\frac{29}{32}$	.9063
$\frac{3}{64}$	.0469	$\frac{11}{64}$	.1719	$\frac{19}{64}$	.2969	$\frac{27}{64}$	.4219	$\frac{35}{64}$	.5469	$\frac{43}{64}$	.6719	$\frac{51}{64}$	.7969	$\frac{59}{64}$	.9219
$\frac{1}{16}$	.0625	$\frac{3}{16}$	.1875	$\frac{5}{16}$	.3125	$\frac{7}{16}$	.4375	$\frac{9}{16}$	.5625	$\frac{11}{16}$	.6875	$\frac{13}{16}$	.8125	$\frac{15}{16}$	.9375
$\frac{5}{64}$	.0781	$\frac{13}{64}$	.2031	$\frac{21}{64}$	.3281	$\frac{29}{64}$	.4531	$\frac{37}{64}$	.5781	$\frac{45}{64}$	.7031	$\frac{53}{64}$	.8281	$\frac{61}{64}$	.9531
$\frac{3}{32}$	.0938	$\frac{7}{32}$	.2188	$\frac{11}{32}$	.3438	$\frac{15}{32}$	.4688	$\frac{19}{32}$	.5938	$\frac{23}{32}$	.7188	$\frac{27}{32}$	.8438	$\frac{31}{32}$	.9688
$\frac{7}{64}$	.1094	$\frac{15}{64}$	.2344	$\frac{23}{64}$	.3594	$\frac{31}{64}$	.4844	$\frac{39}{64}$	.6094	$\frac{47}{64}$	.7344	$\frac{55}{64}$	.8594	$\frac{63}{64}$	.9844
$\frac{1}{8}$	.1250	$\frac{1}{4}$	.2500	$\frac{3}{8}$	.3750	$\frac{1}{2}$	.5000	$\frac{5}{8}$	.6250	$\frac{3}{4}$	.7500	$\frac{7}{8}$	.8750	1	1.0000

INCHES AND FRACTIONS REDUCED TO DECIMALS OF A FOOT.

Inch	1″	2″	3″	4″	5″	6″	7″	8″	9″	10″	11″	Inch
	.000	.083	.167	.250	.333	.417	.500	.583	.667	.750	.833	.917	
$\frac{1}{16}$″	.005	.089	.172	.255	.339	.422	.505	.589	.672	.755	.839	.922	$\frac{1}{16}$″
$\frac{1}{8}$″	.010	.094	.177	.260	.344	.427	.510	.594	.677	.760	.844	.927	$\frac{1}{8}$″
$\frac{3}{16}$″	.016	.099	.182	.266	.349	.432	.516	.599	.682	.766	.849	.932	$\frac{3}{16}$″
$\frac{1}{4}$″	.021	.104	.188	.271	.354	.438	.521	.604	.688	.771	.854	.938	$\frac{1}{4}$″
$\frac{5}{16}$″	.026	.109	.193	.276	.359	.443	.526	.609	.693	.776	.859	.943	$\frac{5}{16}$″
$\frac{3}{8}$″	.031	.115	.198	.281	.365	.448	.531	.615	.698	.781	.865	.948	$\frac{3}{8}$″
$\frac{7}{16}$″	.036	.120	.203	.286	.370	.453	.536	.620	.703	.786	.870	.953	$\frac{7}{16}$″
$\frac{1}{2}$″	.042	.125	.208	.292	.375	.458	.542	.625	.708	.792	.875	.958	$\frac{1}{2}$″
$\frac{9}{16}$″	.047	.130	.214	.297	.380	.464	.547	.630	.714	.797	.880	.964	$\frac{9}{16}$″
$\frac{5}{8}$″	.052	.135	.219	.302	.385	.469	.552	.635	.719	.802	.885	.969	$\frac{5}{8}$″
$\frac{11}{16}$″	.057	.141	.224	.307	.391	.474	.557	.641	.724	.807	.891	.974	$\frac{11}{16}$″
$\frac{3}{4}$″	.063	.146	.229	.313	.396	.479	.563	.646	.729	.813	.896	.979	$\frac{3}{4}$″
$\frac{13}{16}$″	.068	.151	.234	.318	.401	.484	.568	.651	.734	.818	.901	.984	$\frac{13}{16}$″
$\frac{7}{8}$″	.073	.156	.240	.323	.406	.490	.573	.656	.740	.823	.906	.990	$\frac{7}{8}$″
$\frac{15}{16}$″	.078	.161	.245	.328	.411	.495	.578	.661	.745	.828	.911	.995	$\frac{15}{16}$″

INDEX.